Space Exploration
Almanac

Space Exploration Almanac
Volume 2

Rob Nagel

Sarah Hermsen, Project Editor

U·X·L
An imprint of Thomson Gale,
a part of The Thomson Corporation

Detroit • New York • San Francisco • San Diego • New Haven, Conn. • Waterville, Maine • London • Munich

629.4/
Sp9
V. 2

Space Exploration: Almanac
Rob Nagel

Project Editor
Sarah Hermsen

Rights Acquisitions and Management
Ann Taylor

Imaging and Multimedia
Dean Dauphinais, Lezlie Light, Dan Newell

Product Design
Pamela Galbreath

Composition
Evi Seoud

Manufacturing
Rita Wimberley

Library of Congress Cataloging-in-Publication Data
Nagel, Rob.
Space exploration. Almanac / Rob Nagel ; Sarah Hermsen, project editor.
    p. cm. – (Space exploration reference library)
    Includes bibliographical references and index.
    ISBN 0-7876-9209-3 (set hardcover : alk. paper) – ISBN 0-7876-9210-7 (volume 1) – ISBN 0-7876-9211-5 (volume 2)
    1. Astronautics–History–Encyclopedias, Juvenile. 2. Outer space–Exploration–History–Encyclopedias, Juvenile. I. Title. II. Series.
    TL788.N287 2004
    629.4'09–dc22
                                                            2004015823

Contents

Volume 2

Reader's Guide

Fascinating and forbidding, space has drawn the attention of humans since before recorded history. People have looked outward, driven by curiosity about the vast universe that surrounds Earth. Unaware of the meaning of the bright lights in the night sky above them, ancient humans thought they saw patterns, images in the sky of things in the landscape around them.

Slowly, humans came to realize that the lights in the sky had an effect on the workings of the planet around them. They sought to understand the movements of the Sun, the Moon, and the other, brighter objects. They wanted to know how those movements related to the changing seasons and the growth of crops.

Still, for centuries, humans did not understand what lay beyond the boundaries of Earth. In fact, with their limited vision, they saw a limited universe. Ancient astronomers relied on naked-eye observations to chart the positions of stars, planets, and the Sun. In the third century B.C.E., philosophers concluded that Earth was the center of the universe. A few dared to question this prevailing belief. In the face of overwhelming opposition and ridicule, they persisted in trying to understand the truth. This belief ruled human affairs until the scientific revolution of the seventeenth century, when scientists used the newly invented telescope to prove that the Sun is the center of Earth's galaxy.

Over time, with advances in science and technology, ancient beliefs were exposed as false. The universe ever widened with humans' growing understanding of it. The dream to explore its vast reaches passed from nineteenth-century fiction writers to twentieth-century visionaries to present-day engineers and scientists, pilots, and astronauts.

The quest to explore space intensified around the turn of the twentieth century. By that time, astronomers had built better observatories and perfected more powerful telescopes. Increasingly sophisticated technologies led to the discovery that the universe extends far beyond the Milky Way and holds even deeper mysteries, such as limitless galaxies and unexplained phenomena like black holes. Scientists, yearning to solve those mysteries, determined that one way to accomplish this goal was to penetrate space itself.

Even before the twentieth century, people had discussed ways to travel into space. Among them were science fiction writers, whose fantasies inspired the visions of scientists. Science fiction became especially popular in the late nineteenth century, having a direct impact on early twentieth-century rocket engineers who invented the fuel-propellant rocket. Initially developed as a weapon of war, this new projectile could be launched a greater distance than any human-made object in history, and it eventually unlocked the door to space.

From the mid-twentieth century until the turn of the twenty-first century, the fuel-propellant rocket made possible dramatic advances in space exploration. It was used to propel unmanned satellites and manned space capsules, space shuttles, and space stations. It launched an orbiting telescope that sent spectacular images of the universe back to Earth. During this era of intense optimism and innovation, often called the space age, people confidently went forth to conquer the distant regions of space that have intrigued humans since early times. They traveled to the Moon, probed previously uncharted realms, and contemplated trips to Mars.

Overcoming longstanding rivalries, nations embarked on international space ventures. Despite the seemingly unlimited technology at their command, research scientists, engineers, and astronauts encountered political maneuvering, lack of funds, aging spacecraft, and tragic accidents. As the world settled into the twenty-first century, space exploration faced an uncertain future. Yet, the ongoing exploration of space continued to represent the "final frontier" in the last great age of exploration.

Space Exploration: Almanac chronicles the history of space exploration. It is intended as a brief historical overview of humanity's quest to understand and to explore the universe, from those early stargazers to modern interplanetary missions of discovery.

Features

The two-volume *Space Exploration: Almanac* presents, in fourteen chapters, key developments and milestones in the continuing history of space exploration. The focus ranges from ancient views of a Sun-centered universe to the scientific understanding of the laws of planetary motion and gravity, from the launching of the first artificial satellite to be placed in orbit around Earth to current robotic explorations of near and distant planets in the solar system. Also covered is the development of the first telescopes by men such as Hans Lippershey, who called his device a "looker" and thought it would be useful in war, and Galileo Galilei, who built his own device to look at the stars. The work also details the construction of great modern observatories, both on ground and in orbit around Earth, that can peer billions of light-years into space.

Also examined is the development of rocketry, from thirteenth-century Chinese rockets used in warfare to the large multistage Saturn V rocket used to propel the Apollo astronauts to the Moon; the work of theorists and engineers Konstantin Tsiolkovsky, Robert H. Goddard, and others; a discussion of the Cold War and its impact on space exploration; space missions such as the first lunar landing; and great tragedies including the explosions of U.S. space shuttles *Challenger* and *Columbia* as well as the Nedelin catastrophe, in which one hundred Soviet technicians were incinerated as they approached an unstable rocket that had failed to lift off in 1960.

The chapters in *Space Exploration: Almanac* contain sidebar boxes that highlight people and events of special interest, and each chapter offers a list of additional sources that students can go to for more information. More than one hundred black-and-white photographs illustrate the material. Each volume begins with a timeline of important events in the history of space exploration, a "Words to Know" section that introduces students to difficult or unfamiliar terms, and a "Research and Activity Ideas" section. The two volumes conclude with a general bibliography and a subject index so students can easily find the people, places, and events discussed throughout *Space Exploration: Almanac.*

Space Exploration Reference Library

Space Exploration: Almanac is only one component of the three-part Space Exploration Reference Library. The other two titles in this set are:

- *Space Exploration: Biographies* captures the height of the space age in twenty-five entries that profile astronauts, scientists, theorists, writers, and spacecraft. Included are astronauts Neil Armstrong, John Glenn, Mae Jemison, and Sally Ride; cosmonaut Yuri Gagarin; engineer Wernher von Braun; writer H. G. Wells; and the crew of the space shuttle *Challenger.* The volume also contains profiles of the Hubble Space Telescope and the International Space Station. Focusing on international contributions to the quest for knowledge about space, this volume takes readers on an adventure into the achievements and failures experienced by explorers of space.

- *Space Exploration: Primary Sources* (one volume) captures the space age with full-text reprints and lengthy excerpts of seventeen documents that include science fiction, nonfiction, autobiography, official reports, articles, interviews, and speeches. Covering a span of more than one hundred years, these excerpts provide a wide range of perspectives on space exploration, from nineteenth-century speculations about space travel through twenty-first century plans for human flights to Mars. Included are excerpts from science fiction writer Jules Verne's *From the Earth to the Moon;* Tom Wolfe's *The Right Stuff,* which chronicles the story of America's first astronauts; astronaut John Glenn's memoirs; and president George W. Bush's new vision of space exploration.

- A cumulative index of all three titles in the Space Exploration Reference Library is also available.

Comments and Suggestions

We welcome your comments on *Space Exploration: Almanac* and suggestions for other topics to consider. Please write: Editors, *Space Exploration: Almanac,* U•X•L, 27500 Drake Rd. Farmington Hills, Michigan 48331-3535; call toll-free: 1-800-877-4253; fax to (248) 699-8097; or send e-mail via http://www.gale.com.

Timeline of Events

c. 3000 B.C.E. Sumerians produce the oldest known drawings of constellations as recurring designs on seals, vases, and gaming boards.

c. 3000 B.C.E. Construction begins on Stonehenge.

c. 700 B.C.E. Babylonians have already assembled extensive, relatively accurate records of celestial events, including charting the paths of planets and compiling observations of fixed stars.

c. 550 B.C.E. Greek philosopher and mathematician Pythagoras argues that Earth is round and develops an early system of cosmology to explain the nature and structure of the universe.

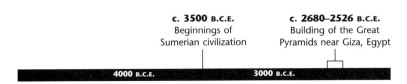

c. 3500 B.C.E.
Beginnings of
Sumerian civilization

c. 2680–2526 B.C.E.
Building of the Great
Pyramids near Giza, Egypt

4000 B.C.E.   3000 B.C.E.

c. 370 B.C.E. Eudoxus of Cnidus develops a system to explain the motions of the planets based on spheres.

c. 280 B.C.E. Greek mathematician and astronomer Aristarchus proposes that the planets, including Earth, revolve around the Sun.

c. 240 B.C.E. Greek astronomer and geographer Eratosthenes calculates the circumference of Earth with remarkable accuracy from the angle of the Sun's rays at separate points on the planet's surface.

c. 130 B.C.E. Greek astronomer Hipparchus develops the first accurate star map and star catalog covering about 850 stars, including a scale of magnitude to indicate the apparent brightness of the stars; it is the first time such a scale has been used.

140 C.E. Alexandrian astronomer Ptolemy publishes his Earth-centered or geocentric theory of the solar system.

c. 1000 The Maya build El Caracol, an observatory, in the city of Chichén Itzá.

1045 A Chinese government official publishes the *Wu-ching Tsung-yao* (*Complete Compendium of Military Classics*), which details the use of "fire arrows" launched by charges of gunpowder, the first true rockets.

1268 English philosopher and scientist Roger Bacon publishes a book on chemistry called *Opus Majus* (*Great Work*) in which he describes in detail the process of making gunpowder, becoming the first European to do so.

1543 Polish astronomer Nicolaus Copernicus publishes his Sun-centered, or heliocentric, theory of the solar system.

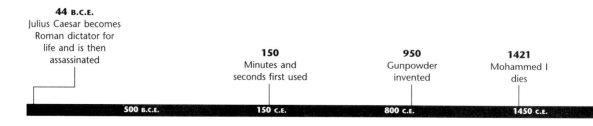

44 B.C.E.
Julius Caesar becomes Roman dictator for life and is then assassinated

150
Minutes and seconds first used

950
Gunpowder invented

1421
Mohammed I dies

500 B.C.E.   **150 C.E.**   **800 C.E.**   **1450 C.E.**

November 1572 Danish astronomer Tycho Brahe discovers what later proves to be a supernova in the constellation of Cassiopeia.

1577 German armorer Leonhart Fronsperger writes a book on firearms in which he describes a device called a *roget* that uses a base of gunpowder wrapped tightly in paper. Historians believe this resulted in the modern word "rocket."

c. late 1500s German fireworks maker Johann Schmidlap invents the step rocket, a primitive version of a multi-stage rocket.

1608 Dutch lens-grinder Hans Lippershey creates the first optical telescope.

1609 German astronomer Johannes Kepler publishes his first two laws of planetary motion.

1609 Italian mathematician and astronomer Galileo Galilei develops his own telescope and uses it to discover four moons around Jupiter, craters on the Moon, and the Milky Way.

1633 Galileo is placed under house arrest for the rest of his life by the Catholic Church for advocating the heliocentric theory of the solar system.

1656 French poet and soldier Savinien de Cyrano de Bergerac publishes a fantasy novel about a man who travels to the Moon in a device powered by exploding firecrackers.

1687 English physicist and mathematician Isaac Newton publishes his three laws of motion and his law of universal gravitation in the much-acclaimed *Philosophiae Naturalis Principia Mathematica* (*Mathematical Principles of Natural Philosophy*).

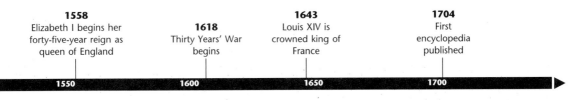

1558
Elizabeth I begins her forty-five-year reign as queen of England

1618
Thirty Years' War begins

1643
Louis XIV is crowned king of France

1704
First encyclopedia published

1550   1600   1650   1700

1781 English astronomer William Herschel discovers the planet Uranus using a reflector telescope he had made.

1804 English artillery expert William Congreve develops the first ship-fired rockets.

1844 English inventor William Hale invents the stickless, spin-stabilized rocket.

1865 French writer Jules Verne publishes *From the Earth to the Moon,* the first of two novels he would write about traveling to the Moon.

1897 The Yerkes Observatory in Williams Bay, Wisconsin, which houses the largest refractor telescope in the world, is completed.

1903 Russian scientist and rocket expert Konstantin Tsiolkovsky publishes an article titled "Exploration of the Universe with Reaction Machines," in which he presents the basic formula that determines how rockets perform.

1923 German physicist Hermann Oberth publishes a ninety-two-page pamphlet titled *Die Rakete zu den Planetenräumen* (*The Rocket into Interplanetary Space*) in which he explains the mathematical theory of rocketry, speculates on the effects of spaceflight on the human body, and theorizes on the possibility of placing satellites in space.

March 16, 1926 American physicist and space pioneer Robert H. Goddard launches the world's first liquid-propellant rocket.

1929 Using the Hooker Telescope at the Mount Wilson Observatory in southern California, U.S. astronomer Edwin Hubble develops what comes to be known as Hubble's law, which describes the rate of expansion of the universe.

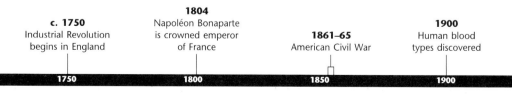

c. 1750
Industrial Revolution begins in England

1804
Napoléon Bonaparte is crowned emperor of France

1861–65
American Civil War

1900
Human blood types discovered

1750   1800   1850   1900

1930   The International Astronomical Union (IAU) sets the definitive boundaries of the eighty-eight recognized constellations.

September 8, 1944 Germany launches V-2 rockets, the first true ballistic missiles, to strike targets in Paris, France, and London, England.

1947   The 200-inch-diameter Hale Telescope becomes operational at the Palomar Observatory in southern California.

March 9, 1955 German-born American engineer Wernher von Braun appears on "Man in Space," the first of three space-related television shows he and American movie producer Walt Disney create for American audiences.

July 1, 1957, to December 31, 1958 During this eighteen-month period, known as the International Geophysical Year, more than ten thousand scientists and technicians representing sixty-seven countries engage in a comprehensive series of global geophysical activities.

October 4, 1957 The Soviet Union launches the world's first artificial satellite, *Sputnik 1,* and the space age begins.

January 31, 1958 *Explorer 1,* the United States's first successful artificial satellite, is launched into space.

March 17, 1958 The U.S. Navy launches the small, artificial satellite *Vanguard 1.* The oldest human-made object in space, it remains in orbit around Earth.

October 1, 1958 The National Aeronautics and Space Administration (NASA) begins work.

January 2, 1959 The Soviet Union launches the space probe *Luna 1,* which becomes the first human-made object to escape Earth's gravity.

1914–18
World War I

1929
Great Depression
begins

1939–45
World War II

1950
Korean War
begins

1920   1930   1940   1950

April 9, 1959 NASA announces the selection of the first American astronauts—the Mercury 7 astronauts: M. Scott Carpenter, Leroy G. "Gordo" Cooper Jr., John Glenn, Virgil I. "Gus" Grissom, Walter M. "Wally" Schirra Jr., Alan B. Shepard Jr., and Donald K. "Deke" Slayton.

September 13, 1959 The Soviet space probe *Luna 2* becomes the first human-made object to land on the Moon when it makes a hard landing east of the Sea of Serenity.

August 18, 1960 The United States launches *Discoverer 14*, its first spy satellite.

October 23, 1960 More than one hundred Soviet technicians are incinerated when a rocket explodes on a launch pad. Known as the Nedelin catastrophe, it is the worst accident in the history of the Soviet space program.

April 12, 1961 Soviet cosmonaut Yuri Gagarin orbits Earth aboard *Vostok 1*, becoming the first human in space.

May 5, 1961 U.S. astronaut Alan Shepard makes a suborbital flight in the capsule *Freedom 7*, becoming the first American to fly into space.

May 25, 1961 U.S. president John F. Kennedy announces the goal to land an American on the Moon by the end of the 1960s.

February 20, 1962 U.S. astronaut John Glenn becomes the first American to circle Earth when he makes three orbits in the *Friendship 7* Mercury spacecraft.

August 27, 1962 *Mariner 2* is launched into orbit, becoming the first interplanetary space probe.

June 16, 1963 Soviet cosmonaut Valentina Tereshkova rides aboard *Vostok 6*, becoming the first woman in space.

1954
Measles vaccine developed

1957
U.S. Congress passes the Civil Rights Act

1961
Bay of Pigs invasion

1964
Supercomputer debuts

1955   1958   1961   1964

November 1, 1963 The world's largest single radio telescope, at Arecibo Observatory in Puerto Rico, officially begins operation.

March 18, 1965 During the Soviet Union's *Voskhod 2* orbital mission, cosmonaut Alexei Leonov performs the first spacewalk, or extravehicular activity (EVA).

February 3, 1966 The Soviet Union's *Luna 9* soft-lands on the Moon and sends back to Earth the first images of the lunar surface.

January 27, 1967 Three U.S. astronauts—Gus Grissom, Roger Chaffee, and Edward White—die of asphyxiation when a fire breaks out in the capsule of *Apollo 1* during a practice session as it sits on the launch pad at Kennedy Space Center, Florida.

April 24, 1967 Soviet cosmonaut Vladimir Komarov becomes the first fatality during an actual spaceflight when the parachute from *Soyuz 1* fails to open and the capsule slams into the ground after reentry.

December 24, 1968 *Apollo 8,* with three U.S. astronauts aboard, becomes the first manned spacecraft to enter orbit around the Moon.

July 20, 1969 U.S. astronauts Neil Armstrong and Buzz Aldrin become the first humans to walk on the Moon.

April 14, 1970 An oxygen tank in the *Apollo 13* service module explodes while the craft is in space, putting the lives of the three U.S. astronauts onboard into serious jeopardy.

December 14, 1970 U.S. astronauts Eugene Cernan and Harrison Schmitt lift off from the Moon after having spent seventy-five hours on the surface. They are the last humans to have set foot on the Moon as of the early twenty-first century.

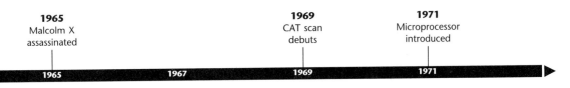

1965
Malcolm X
assassinated

1969
CAT scan
debuts

1971
Microprocessor
introduced

1965   1967   1969   1971

December 15, 1970 The Soviet space probe *Venera 7* arrives at Venus, making the first-ever successful landing on another planet.

April 19, 1971 The Soviet Union launches *Salyut 1,* the first human-made space station.

November 13, 1971 The U.S. probe *Mariner 9* becomes the first spacecraft to orbit another planet when it enters orbit around Mars.

January 5, 1972 U.S. president Richard M. Nixon announces the decision to develop a space shuttle.

May 14, 1973 *Skylab,* the first and only U.S. space station, is launched.

December 4, 1973 The U.S. space probe *Pioneer 10* makes the first flyby of Jupiter.

March 29, 1974 The U.S. space probe *Mariner 10* makes the first of three flybys of Mercury.

July 15 to 24, 1975 The Apollo-Soyuz Test Project is undertaken as an international docking mission between the United States and the Soviet Union.

July 20, 1976 The lander of the U.S. space probe *Viking 1* makes the first successful soft landing on Mars.

September 17, 1976 The first space shuttle orbiter, known as OV-101, rolls out of an assembly facility in Palmdale, California.

January 26, 1978 NASA launches the International Ultraviolet Explorer, considered the most successful UV satellite and perhaps the most productive astronomical telescope ever.

July 11, 1979 *Skylab* falls into Earth's atmosphere and burns up over the Indian Ocean.

1973
U.S. troops pull
out of Vietnam

1977
Star Wars is
released

1978
Test-tube
baby born

1973   1975   1977   1979

October 1979 The United Kingdom Infrared Telescope, the world's largest telescope dedicated solely to infrared astronomy, begins operation in Hawaii near the summit of Mauna Kea.

November 12, 1980 The U.S. probe *Voyager 1* makes a flyby of Saturn and sends back the first detailed photographs of the ringed planet.

April 12, 1981 U.S. astronauts John W. Young and Robert L. Crippen fly the space shuttle *Columbia* on the first orbital flight of NASA's new reusable spacecraft.

June 18, 1983 U.S. astronaut Sally Ride becomes America's first woman in space when she rides aboard the space shuttle *Challenger*.

August 30, 1983 U.S. astronaut Guy Bluford flies aboard the space shuttle *Challenger*, becoming the first African American in space.

January 25, 1984 U.S. president Ronald Reagan directs NASA to develop a permanently manned space station within a decade.

January 28, 1986 The space shuttle *Challenger* explodes seventy-three seconds after launch because of poorly sealing O-rings on the booster rocket, killing all seven astronauts aboard.

February 20, 1986 The Soviet Union launches the core module of its new space station, *Mir,* into orbit.

May 4, 1989 The space shuttle *Atlantis* lifts off carrying the *Magellan* probe, the first planetary explorer to be launched by a space shuttle.

April 25, 1990 Astronauts aboard the space shuttle *Discovery* deploy the Hubble Space Telescope.

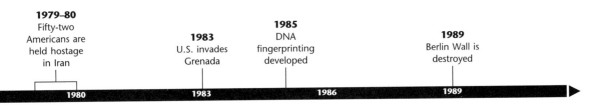

1979–80
Fifty-two Americans are held hostage in Iran

1983
U.S. invades Grenada

1985
DNA fingerprinting developed

1989
Berlin Wall is destroyed

1980   1983   1986   1989

April 7, 1991 The Compton Gamma Ray Observatory is placed into orbit by astronauts aboard the space shuttle *Atlantis.*

December 1993 Astronauts aboard the space shuttle *Endeavour* complete repairs to the primary mirror of the Hubble Space Telescope.

February 3, 1995 The space shuttle *Discovery* lifts off under the control of U.S. astronaut Eileen M. Collins, the first female pilot on a shuttle mission.

December 2, 1995 The Solar and Heliospheric Observatory is launched to study the Sun.

December 7, 1995 The U.S. space probe *Galileo* goes into orbit around Jupiter, dropping a mini-probe to the planet's surface.

March 24, 1996 U.S. astronaut Shannon Lucid begins her 188-day stay aboard *Mir,* a U.S. record for spaceflight endurance at that time.

October 1996 The second of the twin 33-foot Keck telescopes on Mauna Kea, Hawaii, the world's largest optical and infrared telescopes, begins science observations. The first began observations three years earlier.

July 2, 1997 The U.S. space probe *Mars Pathfinder* lands on Mars and releases *Sojourner,* the first Martian rover.

October 15, 1997 The *Cassini-Huygens* spacecraft, bound for Saturn, is launched.

January 6, 1998 NASA launches the *Lunar Prospector* probe to improve understanding of the origin, evolution, current state, and resources of the Moon.

October 29, 1998 At age seventy-seven, U.S. senator John Glenn, one of the original Mercury astronauts, be-

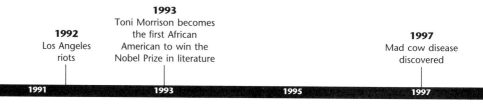

1992
Los Angeles riots

1993
Toni Morrison becomes the first African American to win the Nobel Prize in literature

1997
Mad cow disease discovered

1991　　1993　　1995　　1997

comes the oldest astronaut to fly into space when he lifts off aboard the space shuttle *Discovery.*

November 11, 1998 Russia launches Zarya, the control module and first piece of the International Space Station, into orbit.

July 23, 1999 The Chandra X-ray Observatory is deployed from the space shuttle *Columbia.*

February 21, 2001 The U.S. space probe *NEAR Shoemaker* becomes the first spacecraft to land on an asteroid.

March 23, 2001 After more than 86,000 orbits around Earth, *Mir* enters the atmosphere and breaks up into several large pieces and thousands of smaller ones.

April 28, 2001 U.S. investment banker Dennis Tito, the world's first space tourist, lifts off aboard a Soyuz spacecraft for a week-long stay on the International Space Station.

February 1, 2003 The space shuttle *Columbia* breaks apart in flames above Texas, sixteen minutes before it is supposed to touch down in Florida, because of damage to the shuttle's thermal-protection tiles. All seven astronauts aboard are killed.

June 2003 The Canadian Space Agency launches MOST, its first space telescope successfully launched into space and also the smallest space telescope in the world.

August 25, 2003 NASA launches the Space Infrared Telescope Facility, subsequently renamed the Spitzer Space Telescope, the most sensitive instrument ever to look at the infrared spectrum in the universe.

October 15, 2003 Astronaut Yang Liwei lifts off aboard the spacecraft *Shenzhou 5,* becoming the first Chinese to fly into space.

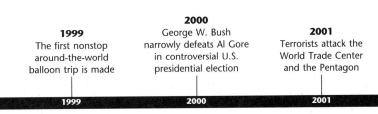

1999
The first nonstop around-the-world balloon trip is made

2000
George W. Bush narrowly defeats Al Gore in controversial U.S. presidential election

2001
Terrorists attack the World Trade Center and the Pentagon

1998   1999   2000   2001

January 14, 2004 U.S. president George W. Bush outlines a new course for U.S. space exploration, including plans to send future manned missions to the Moon and Mars.

June 21, 2004 Civilian pilot Mike Melvill flies the rocket plane *SpaceShipOne* to an altitude of more than 62.5 miles, becoming the first person to pilot a privately built craft beyond the internationally recognized boundary of space.

June 30, 2004 The *Cassini-Huygens* spacecraft becomes the first exploring vehicle to orbit Saturn.

2002
U.S. Justice Department launches investigation into the bankruptcy scandal involving energy giant Enron

2003
The United States declares war on Iraq

| 2002 | 2003 | 2004 |

Words to Know

A

Allies: Alliances of countries in military opposition to another group of nations. In World War II, the Allied powers included Great Britain, the Soviet Union, and the United States.

antimatter: Matter that is exactly the same as normal matter, but with the opposite spin and electrical charge.

apogee: The point in the orbit of an artificial satellite or Moon that is farthest from Earth.

artificial satellite: A human-made device that orbits Earth and other celestial bodies and that follows the same gravitational laws that govern the orbit of a natural satellite.

asterism: A collection of stars within a constellation that forms an apparent pattern.

astrology: The study of the supposed effects of celestial objects on the course of human affairs.

astronautics: The science and technology of spaceflight.

astronomy: The scientific study of the physical universe beyond Earth's atmosphere.

atomic bomb: An explosive device whose violent power is due to the sudden release of energy resulting from the splitting of nuclei of a heavy chemical element (plutonium or uranium), a process called fission.

aurora: A brilliant display of streamers, arcs, or bands of light visible in the night sky, chiefly in the polar regions. It is caused by electrically charged particles from the Sun that are drawn into the atmosphere by Earth's magnetic field.

B

ballistic missile: A missile that travels at a velocity less than what is needed to place it in orbit and that follows a curved path (trajectory) back to Earth's surface once it has reached a given altitude.

bends: A painful and sometimes fatal disorder caused by the formation of gas bubbles in the blood stream and tissues when a decrease in air pressure occurs too rapidly.

big bang theory: The theory that explains the beginning of the universe as a tremendous explosion from a single point that occurred about thirteen billion years ago.

Big Three: The trio of U.S. president Franklin D. Roosevelt, Soviet leader Joseph Stalin, and British prime minister Winston Churchill; also refers to the countries of the United States, the Soviet Union, and Great Britain.

binary star: A pair of stars orbiting around one another, linked by gravity.

black hole: The remains of a massive star that has burned out its nuclear fuel and collapsed under tremendous gravitational force into a single point of infinite mass and gravity from which nothing escapes, not even light.

Bolshevik: A member of the revolutionary political party of Russian workers and peasants that became the Communist Party after the Russian Revolution of 1917.

brown dwarf: A small, cool, dark ball of matter that never completes the process of becoming a star.

C

capitalism: An economic system in which property and businesses are privately owned. Prices, production, and distribution of goods are determined by competition in a market relatively free of government intervention.

celestial mechanics: The scientific study of the influence of gravity on the motions of celestial bodies.

celestial sphere: An imaginary sphere of gigantic radius with Earth located at its center.

Cepheid variable: A pulsating star that can be used to measure distance in space.

chromatic aberration: Blurred coloring of the edge of an image when visible light passes through a lens, caused by the bending of the different wavelengths of the light at different angles.

Cold War: A prolonged conflict for world dominance from 1945 to 1991 between the two superpowers: the democratic, capitalist United States and the Communist Soviet Union. The weapons of conflict were commonly words of propaganda and threats.

Communism: A system of government in which the nation's leaders are selected by a single political party that controls almost all aspects of society. Private ownership of property is eliminated and government directs all economic production. The goods produced and wealth accumulated are, in theory, shared relatively equally by all. All religious practices are banned.

concave lens: A lens with a hollow bowl shape; it is thin in the middle and thick along the edges.

constellation: One of eighty-eight recognized groups of stars that seems to make up a pattern or picture on the celestial sphere.

convex lens: A lens with a bulging surface like the outer surface of a ball; it is thicker in the middle and thinner along the edges.

corona: The outermost and hottest layer of the Sun's atmosphere that extends out into space for millions of miles.

cosmic radiation: High-energy radiation coming from all directions in space.

D

dark matter: Virtually undetectable matter that does not emit or reflect light and that is thought to account for 90 percent of the mass of the universe, acting as a "cosmic glue" that holds together galaxies and clusters of galaxies.

democracy: A system of government that allows multiple political parties. Members of the parties are elected to various government offices by popular vote of the people.

détente: A relaxing of tensions between rival nations, marked by increased diplomatic, commercial, and cultural contact.

docking system: Mechanical and electronic devices that work jointly to bring together and physically link two spacecraft in space.

E

eclipse: The obscuring of one celestial object by another.

ecliptic: The imaginary plane of Earth's orbit around the Sun.

electromagnetic radiation: Radiation that transmits energy through the interaction of electricity and magnetism.

electromagnetic spectrum: The entire range of wavelengths of electromagnetic radiation.

epicycle: A small secondary orbit incorrectly added to the planetary orbits by early astronomers to account for periods in which the planets appeared to move backward with respect to Earth.

escape velocity: The minimum speed that an object, such as a rocket, must have in order to escape completely from the gravitational influence of a planet or a star.

exhaust velocity: The speed at which the exhaust material leaves the nozzle of a rocket engine.

F

flyby: A type of space mission in which the spacecraft passes close to its target but does not enter orbit around it or land on it.

focus: The position at which rays of light from a lens converge to form a sharp image.

force: A push or pull exerted on an object by an outside agent, producing an acceleration that changes the object's state of motion.

G

galaxy: A huge region of space that contains billions of stars, gas, dust, nebulae, and empty space all bound together by gravity.

gamma rays: Short-wavelength, high-energy radiation formed either by the decay of radioactive elements or by nuclear reactions.

geocentric model: The flawed theory that Earth is at the center of the solar system, with the Sun, the Moon, and the other planets revolving around it. Also known as the Ptolemaic model.

geosynchronous orbit: An orbit in which a satellite revolves around Earth at the same rate at which Earth rotates on its axis; thus, the satellite remains positioned over the same location on Earth.

gravity: The force of attraction between objects, the strength of which depends on the mass of each object and the distance between them.

gunpowder: An explosive mixture of charcoal, sulfur, and potassium nitrate.

H

hard landing: The deliberate, destructive impact of a space vehicle on a predetermined celestial object.

heliocentric model: The theory that the Sun is at the center of the solar system and all planets revolve around it. Also known as the Copernican model.

heliosphere: The vast region permeated by charged particles flowing out from the Sun that surrounds the Sun and extends throughout the solar system.

Hellenism: The culture, ideals, and pattern of life of ancient Greece.

hydrocarbon: A compound that contains only two elements, carbon and hydrogen.

hydrogen bomb: A bomb more powerful than the atomic bomb that derives its explosive energy from a nuclear fusion reaction.

hyperbaric chamber: A chamber where air pressure can be carefully controlled; used to acclimate divers, astronauts, and others gradually to changes in air pressure and air composition.

I

inflationary theory: The theory that the universe underwent a period of rapid expansion immediately following the big bang.

infrared radiation: Electromagnetic radiation with wavelengths slightly longer than that of visible light.

interferometer: A device that uses two or more telescopes to observe the same object at the same time in the same wavelength to increase angular resolution.

interplanetary: Between or among planets.

interplanetary medium: The space between planets including forms of energy and dust and gas.

interstellar: Between or among the stars.

interstellar medium: The gas and dust that exists in the space between stars.

ionosphere: That part of Earth's atmosphere that contains a high concentration of particles that have been ionized, or electrically charged, by solar radiation. These particles help reflect certain radio waves over great distances.

J

jettison: To eject or discard.

L

light-year: The distance light travels in the near vacuum of space in one year, about 5.88 trillion miles (9.46 trillion kilometers).

liquid-fuel rocket: A rocket in which both the fuel and the oxidizing agent are in a liquid state.

M

magnetic field: A field of force around the Sun and the planets generated by electrical charges.

magnetism: A natural attractive energy of iron-based materials for other iron-based materials.

magnetosphere: The region of space around a celestial object that is dominated by the object's magnetic field.

mass: The measure of the total amount of matter in an object.

meteorite: A fragment of extraterrestrial material that makes it to the surface of a planet without burning up in the planet's atmosphere.

microgravity: A state where gravity is reduced to almost negligible levels, such as during spaceflight; commonly called weightlessness.

micrometeorite: A very small meteorite or meteoritic particle with a diameter less than a 0.04 inch (1 millimeter).

microwaves: Electromagnetic radiation with a wavelength longer than infrared radiation but shorter than radio waves.

moonlet: A small artificial or natural satellite.

N

natural science: A science, such as biology, chemistry, or physics, that deals with the objects, occurrences, or laws of nature.

neutron star: The extremely dense, compact, neutron-filled remains of a star following a supernova.

nuclear fusion: The merging of two hydrogen nuclei into one helium nucleus, accompanied by a tremendous release of energy.

O

observatory: A structure designed and equipped to observe astronomical phenomena.

oxidizing agent: A substance that can readily burn or promote the burning of any flammable material.

ozone layer: An atmospheric layer that contains a high proportion of ozone molecules that absorb incoming ultraviolet radiation.

P

payload: Any cargo launched aboard a spacecraft, including astronauts, instruments, and equipment.

perigee: The point in the orbit of an artificial satellite or Moon that is nearest to Earth.

physical science: Any of the sciences—such as astronomy, chemistry, geology, and physics—that deal mainly with nonliving matter and energy.

precession: The small wobbling motion Earth makes about its axis as it spins.

probe: An unmanned spacecraft sent to explore the Moon, other celestial bodies, or outer space; some probes are programmed to return to Earth while others are not.

propellant: The chemical mixture burned to produce thrust in rockets.

pulsar: A rapidly spinning, blinking neutron star.

Q

quasars: Extremely bright, star-like sources of radio waves that are found in remote areas of space and that are the oldest known objects in the universe.

R

radiation: The emission and movement of waves of atomic particles through space or other media.

radio waves: The longest form of electromagnetic radiation, measuring up to 6 miles (9.7 kilometers) from peak to peak in the wave.

Red Scare: A great fear among U.S. citizens in the late 1940s and early 1950s that communist influences were infiltrating U.S. society and government and could eventually lead to the overthrow of the American democratic system.

redshift: The shift of an object's light spectrum toward the red end of the visible light range, which is an indication that the object is moving away from the observer.

reflector telescope: A telescope that directs light from an opening at one end to a concave mirror at the far end, which reflects the light back to a smaller mirror that directs it to an eyepiece on the side of the telescope.

refractor telescope: A telescope that directs light waves through a convex lens (the objective lens), which bends the waves and brings them to a focus at a concave lens (the eyepiece) that acts as a magnifying glass.

retrofire: The firing of a spacecraft's engine in the direction opposite to which the spacecraft is moving in order to cut its orbital speed.

rover: A remote-controlled robotic vehicle.

S

sidereal day: The time for one complete rotation of Earth on its axis relative to a particular star.

soft landing: The slow-speed landing of a space vehicle on a celestial object to avoid damage to or the destruction of the vehicle.

solar arrays: Groups of solar cells or other solar collectors arranged to capture energy from the Sun and use it to generate electrical power.

solar day: The average time span from one noon to the next.

solar flare: Temporary bright spot that explodes on the Sun's surface, releasing an incredible amount of energy.

solar prominence: A tongue-like cloud of flaming gas projecting outward from the Sun's surface.

solar wind: Electrically charged subatomic particles that flow out from the Sun.

solid-fuel rocket: A rocket in which the fuel and the oxidizing agent exist in a solid state.

solstice: Either of the two times during the year when the Sun, as seen from Earth, is farthest north or south of the equator; the solstices mark the beginning of the summer and winter seasons.

space motion sickness: A condition similar to ordinary travel sickness, with symptoms that include loss of appetite, nausea, vomiting, gastrointestinal disturbances, and fatigue. The precise cause of the condition is not fully understood, though most scientists agree the problem originates in the balance organs of the inner ear.

space shuttle: A reusable winged spacecraft that transports astronauts and equipment into space and back.

space station: A large orbiting structure designed for long-term human habitation in space.

spacewalk: Technically known as an EVA, or extravehicular activity, an excursion outside a spacecraft or space station by an astronaut or cosmonaut wearing only a pressurized spacesuit and, possibly, some sort of maneuvering device.

spectrograph: A device that separates light by wavelengths to produce a spectrum.

splashdown: The landing of a manned spacecraft in the ocean.

star: A hot, roughly spherical ball of gas that emits light and other forms of electromagnetic radiation as a result of nuclear fusion reactions in its core.

stellar scintillation: The apparent twinkling of a star caused by the refraction of the star's light as it passes through Earth's atmosphere.

stellar wind: Electrically charged subatomic particles that flow out from a star (like the solar wind, but from a star other than the Sun).

sunspot: A cool area of magnetic disturbance that forms a dark blemish on the surface of the Sun.

supernova: The massive explosion of a relatively large star at the end of its lifetime.

T

telescope: An instrument that gathers light or some other form of electromagnetic radiation emitted by distant sources, such as celestial bodies, and brings it to a focus.

thrust: The forward force generated by a rocket.

U

ultraviolet radiation: Electromagnetic radiation of a wavelength just shorter than the violet (shortest wavelength) end of the visible light spectrum.

United Nations: An international organization, composed of most of the nations of the world, created in 1945 to preserve world peace and security.

V

Van Allen belts: Two doughnut-shaped belts of high-energy charged particles trapped in Earth's magnetic field.

X

X rays: Electromagnetic radiation of a wavelength just shorter than ultraviolet radiation but longer than gamma rays that can penetrate solids and produce an electrical charge in gases.

Y

Yalta Conference: A 1944 meeting between Allied leaders Joseph Stalin, Winston Churchill, and Franklin D. Roosevelt in anticipation of an Allied victory in Europe over the Nazis during World War II (1939–45). The leaders discussed how to manage lands conquered by Germany, and Roosevelt and Churchill urged Stalin to enter the Soviet Union in the war against Japan.

Research and Activity Ideas

The following ideas and projects are intended to offer suggestions for complementing your classroom work on understanding various aspects of the history of space exploration:

- **Inventing Constellation Stories:** Pick five modern constellations. Instead of the accepted images associated with those constellations, develop five new ones. The characters or objects could be modern or ancient, real or imagined. Create and write mythologies or stories about those new characters or objects.

- **Becoming Collins:** In July 1969, Michael Collins circled the Moon for more than twenty-four hours alone in the *Apollo 11* command module while his fellow U.S. astronauts, Neil Armstrong and Buzz Aldrin, walked on the Moon. Research to find out what tasks Collins had to perform during his time alone in space. Then pretend you are Collins and write a journal entry for that time, recording your actions. Be sure to include your thoughts and observations about the experience, imagining what you see out of the module's windows and what you are thinking about on your lonely voyage around the Moon.

- **Creating a Space Cartoon:** Using imagination and artistic skills, create a newspaper cartoon about the flight of the first artificial satellite, *Sputnik 1,* or the first manned spaceflight of Soviet cosmonaut Yuri Gagarin. Before beginning the cartoon, determine whether it will appear in a Soviet or U.S. newspaper at the time. Remember that both events occurred during the height of the Cold War when both nations were trying to prove their superiority. Be sure to convey an emotion such as pride, fear, or surprise. Write a caption for the cartoon that captures the essential message or spirit of the cartoon.

- **Recording Oral Histories:** Interview an individual, such as a relative or an acquaintance, who lived during the late 1950s and early 1960s. Find out what they thought about the early space race and the development of space exploration. Did their expectations come to pass? Develop questions ahead of time. Tape record the interview if possible or take careful notes. Transcribe the tapes or rewrite the notes into a clearly written story retelling the interview.

- **Reporting on the Lunar Landing:** Research and read newspaper accounts of the first landing of humans on the Moon. Adopting the persona of a reporter, write an article of the event that would appear in your local newspaper.

- **Sending Animals into Space:** Find out about the animals used in the early days of the Soviet and U.S. space programs. What kinds of animals were sent into space? What happened on their missions and what was learned that later helped manned missions sent into space? Write about your findings in a science article.

- **Using New Products from the Space Age:** Products developed during the Apollo and later NASA projects are now common in daily life. From freeze-dried foods to cordless power tools, many of these have made life on Earth more convenient and comfortable. Research five commonly used products that were developed during the space program. Prepare a display showing how each product was used originally and how each one is used now.

- **Dodging Space Junk:** The exploration of space has resulted not only in great discoveries and triumphs, but has left much "junk" floating in space, especially around

Earth. Research to find out how much and what types of space junk are in orbit around the planet, then write a humorous story about the adventure of circling Earth in a spacecraft while trying to avoid all the junk.

• **Dieting in Space:** Find out what types of food are served aboard the space shuttle and the International Space Station. Using a computer, create a database file. Design a database template that includes fields such as day (1, 2, 3, etc.), meal (breakfast, lunch, dinner, and a possible snack), and the six major food groups (grain, vegetable, fruit, dairy, meat, and fats). Enter the information from the menus and determine which meals are balanced ones by searching for any empty fields in the food groups. Write a short report based on your findings, answering the following questions: Which food groups had the better selection of foods? Why is it important to maintain good health in space? How does a balanced diet promote good health?

• **Debating the Future of Space Exploration:** With other students, form two or three groups and debate the future direction of NASA. Have each group take a different position on issues such as: Should the space shuttle be scrapped? If so, what, if anything, should replace it? Should the United States retain a presence on the International Space Station? Should the United States undertake voyages to the Moon and Mars? What should happen to space-based observatories such as the Hubble Space Telescope?

Space Exploration
Almanac

9

Apollo-Soyuz Test Project

The mid-1950s was a time marked by both scientific and political fervor. The eighteen-month period between July 1, 1957, and December 31, 1958, was known as the International Geophysical Year (IGY). During this period, international cooperation in science peaked. More than ten thousand scientists and technicians representing sixty-seven countries participated in a multitude of cooperative research programs and activities aimed at gathering data about Earth, its atmosphere, and the Sun. Both the United States and the Soviet Union (present-day Russia) used the focus on upper-atmosphere research during the IGY to develop orbiting artificial satellites. Out of this focus would come each country's eventual space program.

While there was cooperation in the scientific world, there was conflict in the political one. World politics was dominated by the differing political ideologies (set of doctrines or beliefs) of the democratic, capitalist United States and the Communist Soviet Union. The mistrust between these two extremely powerful nations had grown after the end of World War II (1939–45). It gave rise to an atmosphere of hostility and fear

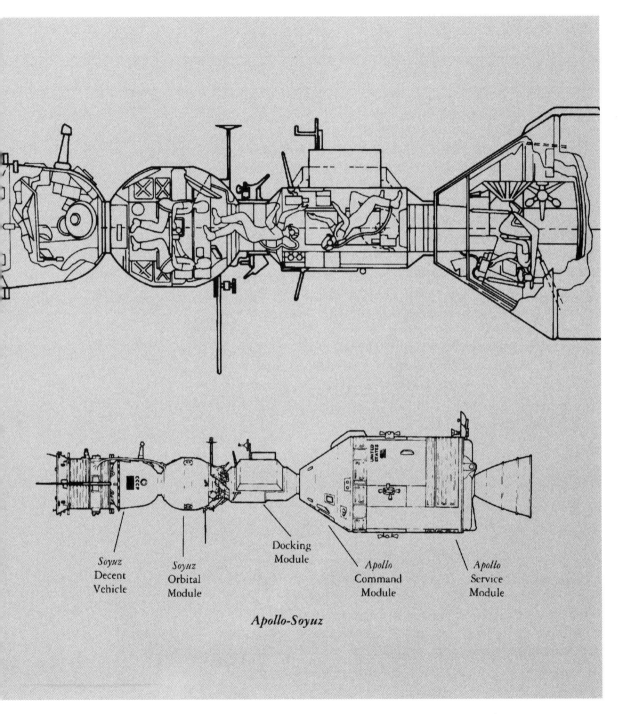

Soyuz
Decent
Vehicle

Soyuz
Orbital
Module

Docking
Module

Apollo
Command
Module

Apollo
Service
Module

Apollo-Soyuz

Various sections of the Apollo-Soyuz spacecrafts, showing astronauts positioned inside. *(National Aeronautics and Space Administration)*

that enveloped the planet. Virtually every significant event or development in world affairs—political, military, economic, and cultural—was directed in response to the unstable relationship between the superpowers. The relationship became a war, mostly of words, about global domination and global destruction. Known as the Cold War, it would last for almost half a century.

Against the background of the Cold War, the United States and the Soviet Union (present-day Russia) engaged in a seemingly bitter race for supremacy in space for more than twelve years. That race began in earnest when the Soviet Union launched the first unmanned artificial satellite, *Sputnik 1,* into orbit around Earth on October 4, 1957. The United States's first success came early the following year when *Explorer 1* was launched on February 1, 1958.

Over the next few years, the Soviets continued to set new space records: Six more Sputniks, each one larger than the first, were launched into Earth's orbit between 1958 and 1961. In 1959 the Soviets sent three Luna probes to the Moon: *Luna 1* was the first probe to fly past the Moon, *Luna 2* was the first to hit the Moon, and *Luna 3* was the first to photograph the Moon's far side. Then on April 12, 1961, cosmonaut Yuri A. Gagarin (1934–1968) flew aboard *Vostok 1*. His historic 108-minute flight marked the first time a human had traveled in space. The United States caught up a month later when Alan Shepard Jr. (1923–2001) made a suborbital (less than a full orbit) flight in the Mercury capsule *Friendship 7*.

Once manned spaceflight had been achieved, the Soviet Union and the United States battled to be the first to put a man on the Moon. In this part of the race, the United States was not in a position where it had to play catch-up: Neither country initially had a rocket powerful enough to complete such a mission. To get to the Moon required, among other technological advances, the development of a super-rocket. In this, the United States succeeded with the Saturn V. The Soviet Union's N-1 never rose above Earth's atmosphere.

As millions watched on television, two U.S. astronauts, Neil Armstrong (1930–) and Edwin E. "Buzz" Aldrin Jr. (1930–), stepped onto the surface of the Moon on July 20, 1969. Many back on Earth believed that this moment, when humans first set foot on another celestial body, signaled the end of the

Words to Know

Artificial satellite: A man-made device that orbits Earth and other celestial bodies and that follows the same gravitational laws that govern the orbit of a natural satellite.

Bends: A painful and sometimes fatal disorder caused by the formation of gas bubbles in the bloodstream and tissues when a decrease in air pressure occurs too rapidly.

Cold War: A prolonged conflict for world dominance from 1945 to 1991 between the democratic, capitalist United States and the Communist Soviet Union. The weapons of conflict were commonly words of propaganda and threats.

Corona: The outermost and hottest layer of the Sun's atmosphere that extends out into space for millions of miles.

Détente: A relaxing of tensions between rival nations, marked by increased diplomatic, commercial, and cultural contact.

Docking system: Mechanical and electronic devices that work jointly to bring together and physically link two spacecraft in space.

Hyperbaric chamber: A chamber where air pressure can be carefully controlled; used to acclimate divers, astronauts, and others gradually to changes in air pressure and air composition.

Solar arrays: Groups of solar cells or other solar collectors arranged to capture energy from the Sun and use it to generate electrical power.

Space shuttle: A reusable winged spacecraft that transports astronauts and equipment into space and back.

Space station: A large orbiting structure designed for long-term human habitation in space.

Spacewalk: Technically known as an EVA, or extravehicular activity, an excursion outside a spacecraft or space station by an astronaut or cosmonaut wearing only a pressurized spacesuit and, possibly, some sort of maneuvering device.

United Nations: An international organization, composed of most of the nations of the world, created in 1945 to preserve world peace and security.

space race with the United States emerging victorious. Others believed that victory had already been sealed when *Apollo 8* went into orbit around the Moon in December 1968, demonstrating that the distance between Earth and the Moon could

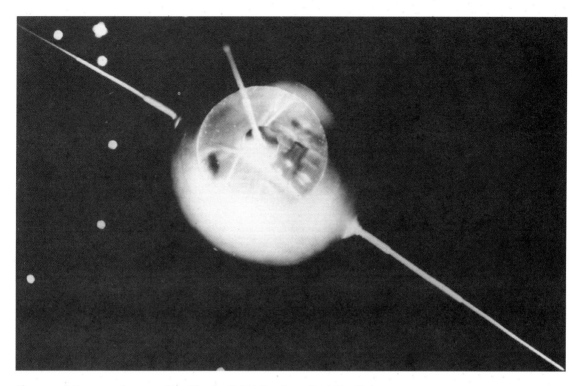

The race into space began when the Soviet Union launched the first unmanned artificial satellite, *Sputnik 1,* into orbit around Earth on **October 4, 1957.** *(AP/Wide World Photos)*

be easily navigated by humans. These incredible achievements in human history, though, had come at a price: Beyond the millions of dollars spent by both sides, seven astronauts and at least two cosmonauts died in training exercises or on missions. (Because of the secrecy in which the Soviets conducted their space program, it is unknown exactly how many early cosmonauts may have perished.)

After the success of Project Apollo (the U.S. lunar-landing program), the National Aeronautics and Space Administration (NASA) faced an uncertain future. The Moon had been conquered, but manned missions to other bodies in the solar system, such as Mars, were decades away, if they were even possible. Plans for reusable space shuttles and an orbiting space station had been developed, but public interest in space exploration had dwindled. The space race had been more political than scientific, and the U.S. public has never had great enthusiasm for purely scientific endeavors. Instead, public attention was focused on rising social problems and an increasingly unpopular war in Vietnam. Due to poor economic conditions in the country, NASA's funding declined, severely limiting the activities and projects it could undertake.

The Soviets, meanwhile, carried out a hoax after they lost the race to the Moon. They claimed they had never been in the race to begin with. The Soviet manned lunar programs were made public only after the breakup of the Soviet Union in the early 1990s. The Soviets had two huge secret projects designed to win the moon race. The aim of the L-1 project was to send a Soviet crew around the Moon before a U.S. crew, using a stripped-down Soyuz spacecraft. The aim of the L-3 project was to land a crew of cosmonauts on the Moon before a crew of U.S. astronauts. Because of equipment malfunctions, both projects were failures.

After quietly giving up their quest to put a cosmonaut on the Moon, the Soviets shifted their focus to other types of space exploration. This included the development of a series of space stations. On April 19, 1971, they launched *Salyut 1,* the world's first space station. Three days later, *Soyuz 10* was launched on the first mission to the station. The spacecraft, which carried three cosmonauts, was unable to dock with *Salyut 1* for an unexplained reason and returned to Earth. Another attempt to dock was then made by *Soyuz 11,* which lifted off on June 6, 1971. After successfully docking, the three cosmonauts—Georgi Dobrovolski (1928–1971), Vladislav Volkov (1935–1971), and Viktor Patsayev (1933–1971)—spent twenty-four days testing the space station's systems and conducting biomedical and other scientific work. Then tragedy struck. As *Soyuz 11,* undocked from *Salyut 1,* a valve on the spacecraft that was supposed to equalize pressure inside the capsule in the final moments before landing was jolted open. It remained open during reentry, allowing all of the air in the capsule to escape. When Soviet officials opened the capsule after it had landed, they found that all three cosmonauts had suffocated to death.

After this disaster, the Soviets decided to scrap *Salyut 1.* They programmed it to reenter Earth's atmosphere on October 11, 1971. Six months after it had lifted off into space, the space station was incinerated.

The Soviet *Soyuz 11* rocket successfully launched, but tragically all three cosmonauts onboard died when a valve jolted open and the spacecraft lost cabin pressure. *(AP/Wide World Photos)*

United Nations Office for Outer Space Affairs

The United Nations (UN) office responsible for promoting international cooperation in the peaceful uses of outer space is known as the United Nations Office for Outer Space Affairs (UNOOSA). It serves as the secretariat or department for the Committee on the Peaceful Uses of Outer Space, the UN's only committee dealing exclusively with international cooperation in the peaceful uses of outer space. In 2004 there were sixty-five member states in the committee, making it one of the largest committees in the UN.

UNOOSA, located in the UN office in Vienna, Austria, prepares and distributes reports and publications on various fields of space science, technology applications, and international space law. It works to improve the use of space science and technology for the economic and social development of all nations, particularly developing countries. It also maintains the Register of Objects Launched into Outer Space.

As both the Soviet and U.S. space programs set about recovering from tragedy and a loss of direction, the idea of setting aside political differences and cooperating on the exploration of space seemed proper and necessary. It was not a new idea. In 1959 the United Nations (UN; an international organization, composed of most of the nations of the world, created in 1945 to preserve world peace and security) had set up a permanent committee to review and encourage international cooperation in the peaceful uses of outer space and to study legal problems arising from its exploration. It is known as the Committee on the Peaceful Uses of Outer Space.

Certain officials from each of the competing space programs desired closer relations, but the Soviet and U.S. governments remained highly suspicious of each other. The political walls between them remained high. Nonetheless, communication between NASA and its Soviet counterpart continued throughout the 1960s. Then a thaw came.

Détente

From 1969 through 1975, the United States and the Soviet Union established policies promoting détente between them. Détente (pronounced day-TONT; French for "lessening of tensions") marked a relaxing of tensions between the rival nations, represented by increased diplomatic, commercial, and cultural contact. It emerged as part of the cyclical pattern of Cold War history, in which periods of relative calm followed periods of bitter superpower conflict. Western and Eastern European countries also experienced a détente and better cooperation during this period.

Consistent contact and communication between the United States and the Soviet Union was perhaps the single greatest achievement of détente. The détente period is also significant because it marked the beginning of improved relations between the United States and Communist China. Recognizing that China and the United States could become allies pushed the Soviets toward détente. These positive changes were bright spots at a time when the United States seemed all but consumed by the challenge of removing itself from the Vietnam War (1954–75).

Vietnam had come under French colonial rule in the late nineteenth century. In 1954 Communist forces led by Ho Chi Minh (1890–1969) defeated the French and Vietnam was then temporarily divided in two, pending general elections to bring about national reunification. North Vietnam continued under Communist leadership, while South Vietnam aligned itself with the United States. Soon, Communist forces from North Vietnam (Viet Cong) launched attacks against the South with the purpose of unifying the two Vietnams under Communist rule. U.S. president Dwight D. Eisenhower (1890–1969) propped up the South Vietnamese government with substantial economic and military aid. His successor as president, John F. Kennedy (1917–1963), expanded that commitment, broadening the U.S. military role because he believed that Ho Chi Minh and his forces were part of a general Communist expansion around the world. Like most Americans, he believed that the Communist governments of the Soviet Union and China controlled North Vietnam. Kennedy vowed not to lose Vietnam to the Communists. Beginning in early 1965, U.S. combat troops were introduced in growing numbers into Vietnam, and the war escalated.

By 1970 the Vietnam War had become the single greatest political controversy in the United States. The war, supported by very few U.S. international allies, had eroded confidence in U.S. power at home and abroad. The enormous financial and human cost of the war to the United States (more than 110 billion dollars at the time and more than 58,000 lives and 300,000 casualties) jeopardized the readiness of U.S. military forces. The huge expense of the war fueled inflation (the continuing rise in the general price of goods and services because of an overabundance of available money) and threatened to send the nation into a recession (a period of extended economic decline). In the United States, opposition to the war increased steadily, dividing the public and straining the relationship between U.S. president Richard M. Nixon (1913–1994) and the U.S. Congress.

President Nixon had been elected in 1968 in part because he hinted that he had a plan to end the war and withdraw U.S. troops from Vietnam. In fact, he had no such plan. Nixon considered immediate withdrawal from Vietnam impossible. Such a drastic move might trigger a political backlash from U.S. supporters of the war, impairing Nixon's ability to draft domestic legislation and negotiate with foreign powers, especially the Soviet Union.

Nixon was keenly interested in improving relations between the United States and the Communist powers of the Soviet Union and China. However, he did not take the lead in détente. That movement actually came from Europe, spurred on by French president Charles de Gaulle (1890–1970) and West German chancellor Willy Brandt (1913–1992). Disgusted by the war in Vietnam, de Gaulle condemned the United States as a reckless world power. Both he and Brandt sought to open communications with the governments of Eastern Europe and the Soviet Union. In doing so, they achieved major improvements in East-West relations.

In part to prevent the Europeans from undermining the United States's leadership in the world, Nixon and his national security advisor, Henry A. Kissinger (1923–), pursued détente. They also hoped to use improved relations to gain the assistance of the Soviet Union and China in bringing an end to the widely unpopular Vietnam War without a humiliating defeat for the United States.

The first cooperative venture in space

Détente opened the door for a cooperative space mission. In March 1970 President Nixon declared international cooperation a prime objective of NASA. On October 24 of that year, a U.S. delegation led by NASA deputy administrator George Low (1926–1984) met with Soviet officials in Moscow (the Soviet capital) to begin talks on the development of a common docking system that would allow each country to rescue the other's space travelers. Negotiations for a joint endeavor in space continued over the next nineteen months until Nixon made a highly publicized visit to Moscow in May 1972. The primary purpose of the trip was to sign the Strategic Arms

U.S. president Richard Nixon, at left, and Soviet leader Leonid Brezhnev exchange copies of the signed nuclear arms race treaty, 1972. *(© Bettmann/Corbis)*

Limitation Treaty (SALT), which was designed to slow the nuclear arms race, limiting the number of offensive missiles each side had. As part of that accord, Nixon and Soviet premier Alexei Kosygin (1904–1980) signed the "Agreement Concerning Cooperation in the Exploration and Use of Outer Space for Peaceful Purposes." In addition to the first in-orbit manned space mission, the five-year agreement called for a wide range of continuing cooperative activities in such areas as space meteorology, space biology and medicine, the study of the natural environment from space, and the exploration of near-Earth space.

That mission, known officially as the Apollo-Soyuz Test Project (ASTP; the Soviets referred to it as the Soyuz-Apollo Test Project), was designed mainly to develop space-based rescue techniques needed by both manned space programs. Science experiments would be conducted in space once the two spacecraft had docked. Mastering the logistics of such a joint space operation would help pave the way for future joint ventures involving space shuttles and space stations.

On January 30, 1973, NASA introduced the crew that would fly the ASTP Apollo spacecraft (unofficially designated *Apollo 18*): Thomas P. Stafford (1930–), Vance Brand (1931–), and Donald "Deke" Slayton (1924–1993). Stafford was a veteran astronaut who had flown aboard *Gemini 6, Gemini 9,* and *Apollo 7.* Brand was a rookie astronaut, having flown on no previous missions. While Slayton had never flown on a space mission, he was hardly a novice. In 1959 he had been selected as one of the original seven Mercury astronauts. He had been assigned to fly the second Mercury orbital mission, but NASA doctors discovered that he had an irregular heartbeat and grounded him. He stayed with NASA to supervise the astronaut corps, eventually as director of flight crew operations. It was his job to decide who would fly aboard missions into space. In 1972, after having overcome his heart problem, he was restored to flight status. When Slayton finally flew into space, he had been an astronaut for sixteen years.

Five months after the Apollo crew had been chosen, the Soviets announced that cosmonauts Aleksei Leonov (1934–) and Valeri Kubasov (1935–) would fly aboard *Soyuz 19.* Leonov already had a marked history in spaceflight: In March 1965, while orbiting Earth aboard *Voskhod 2,* he performed the first spacewalk, or EVA (extravehicular activity). He had also been selected to command the first Soviet manned mission to orbit the Moon, had it been attempted. Kubasov had flown in space as a member of the *Soyuz 6* crew in 1969.

For the mission, the two crews trained together in Houston, Texas, and in Moscow. Part of the training involved learning each other's language. This proved a bit difficult. The astronauts and cosmonauts agreed to talk to their respective mission controllers in their native language; communication between the crews would consist of simple words, gestures, and sign language. The Houston and Moscow mission control centers also learned to work together. Meanwhile, U.S. and Soviet engineers worked to make the ASTP spacecraft compatible.

The Apollo-Soyuz crew (left to right): Donald Slayton, Thomas Stafford, Vance Brand, Aleksei Leonov, and Valeri Kubasov.
(AP/Wide World Photos)

Both a common docking unit and a Docking Module (DM) had to be built. The docking unit, the Androgynous Peripheral Docking System (APDS), was based on a U.S. design. Unlike previous docking units used in space, the APDS could play both passive and active roles in docking. In the active role, motors on the APDS both extended and retracted the unit. Spade-shaped guides aligned the APDS units on the two spacecraft so latches could hook them together. In the Apollo APDS, shock absorbers softened the impact; on the Soyuz APDS, a gear system performed the same function. Once the ships were docked, the active APDS then retracted to lock the ships together and create an airtight seal.

The DM, built by the United States, allowed movement between the incompatible Apollo and Soyuz atmospheres by acting as a hyperbaric chamber, a compartment where air pressure can be carefully controlled. Apollo had a low-pressure, high-oxygen atmosphere, whereas Soyuz had an atmosphere that replicated Earth's (an oxygen-nitrogen mixture at three time the pressure in Apollo). If the cosmonauts had transferred directly from *Soyuz 19* to the Apollo capsule without the DM, they would have suffered from the bends, a painful and sometimes fatal disorder caused by the formation of gas bubbles in the blood stream and tissues when a decrease in air pressure occurs too rapidly.

The 4,432-pound (2,010-kilogram) DM included an Apollo-type docking unit on one end and the APDS docking unit on the other. It was launched underneath the Apollo spacecraft, on top of the two-stage Saturn IB rocket. (Apollo missions to the Moon were launched on larger Saturn V rockets.) The ASTP Apollo spacecraft was a stripped-down Apollo lunar spacecraft. In keeping with its short-duration, Earth-orbital mission, it carried few supplies and little propellant. At roughly 28,000 pounds (12,710 kilograms), it was the lightest Apollo spacecraft ever flown.

Modifications to the Soyuz spacecraft included replacing the standard Soyuz docking system (designed for docking with Salyut space stations) with the Soviet APDS, adding electricity-generating solar arrays (groups of solar cells or other solar collectors arranged to capture energy from the Sun and use it to generate electrical power), and making upgrades to the life-support systems so the cosmonauts could host the visiting astronauts.

The historic mission

In the time previous to this historic international space mission, the United States had sent thirty-four men into space on the twenty-seven manned missions comprising the Mercury, Gemini, and Apollo programs. A number of those astronauts flew on two or more missions. Only one astronaut, Walter Schirra (1935–), flew in all three programs. The Soviet Union had sent thirty-three men and one woman into space on the twenty-six missions comprising its Vostok, Voskhod, and Soyuz programs. Like their U.S. counterparts, a number of the cosmonauts flew on more than one mission. However, no cosmonaut flew in all three programs.

On July 15, 1975, *Soyuz 19* lifted off flawlessly from a launch pad at the Baikonur Space Center in the desert in present-day Kazakhstan. About seven hours later, ASTP Apollo lifted off from Kennedy Space Center in Cape Canaveral, Florida. Shortly after achieving orbit, the Apollo spacecraft turned around and docked with the DM in the separated second stage of the Saturn IB. The astronauts then maneuvered the spacecraft, which was traveling at a speed of approximately 17,480 miles (28,125 kilometers) per hour, into a proper orbit so it could rendezvous with *Soyuz 19*.

After chasing down the Soyuz spacecraft, ASTP Apollo rendezvoused and docked with it at 12:10 P.M. EDT on July 17. The docking, in which the Apollo APDS played the active role, was flawless. Stafford and Slayton then entered the DM, closing the hatch to the Apollo spacecraft behind them. They adjusted the air pressure inside, raising it to meet the cabin pressure inside *Soyuz 19*, which the cosmonauts had lowered slightly before docking. As the joined spacecraft passed over the French city of Metz, Stafford opened the hatch that led into the Soyuz spacecraft. With applause from both mission control centers in the background, the two commanders, Stafford and Leonov, shook hands. It was an event that was broadcast live on global television.

Over the next two days, the two crews conducted four transfers between their spacecraft. During these, much attention was given to television coverage and symbolism. The astronauts and cosmonauts shared a meal, heard greetings from U.S. president Gerald R. Ford (1913–) and Soviet premier Leonid Brezhnev (1906–1982), and exchanged plaques, flags, certificates, and other gifts. Leonov and Kubasov gave the U.S. public a television tour of their spacecraft, and the astronauts did the same for a Soviet audience. Although science was of secondary importance during the mission, the crews did conduct twenty-seven experiments during the entire mission, some jointly and the rest independently.

After having been docked for forty-four hours, the two spacecraft separated. The ASTP Apollo then maneuvered between *Soyuz 19* and the Sun, creating an artificial solar eclipse

Ceremonial Items Carried Aboard the ASTP Mission

The Apollo-Soyuz Test Project (ASTP) was the first cooperative international manned spaceflight. To mark the historic event, the U.S. astronauts aboard the Apollo spacecraft and the Soviet cosmonauts aboard the Soyuz spacecraft carried into space and exchanged a number of items, including:

U.S. and U.S.S.R. flags (for exchange): The astronauts and cosmonauts carried five small flags from their respective nations that they exchanged with each other. The national flags symbolized the contribution made by many people from across the United States and Soviet Union.

U.S. and U.S.S.R. flags (not for exchange): The astronauts and cosmonauts also carried one large flag and five small flags from their respective nations that they did not exchange, but retained to symbolize the role that each nation played in the first international manned spaceflight.

United Nations flag: The cosmonauts carried aloft a large United Nations flag, which was then returned to Earth by the astronauts. It symbolized the contribution to this and other cooperative space projects made by people from many nations and the common goal of exploring space peacefully for the benefit of all people.

Commemorative medallions: The astronauts and cosmonauts each carried two halves of two individual medallions with crossed flags and docked spacecraft. They

when viewed from the Soyuz spacecraft. This allowed the cosmonauts to photograph the Sun's corona, the outermost and hottest layer of the solar atmosphere that extends out into space for millions of miles. After this experiment, the Apollo spacecraft moved toward *Soyuz 19* for redocking. This time, the Soyuz spacecraft was the prime maneuvering vehicle and its APDS the active docking unit. Once again, docking was successful. After three hours, the spacecraft undocked for a second and final time. After conducting a few more joint experiments, ASTP Apollo and *Soyuz 19* went their separate ways.

The Soviet craft remained in orbit for one-and-a-half more days, conducting experiments. It then reentered Earth's atmosphere on July 21, landing safely in present-day Kazakhstan.

exchanged one of the halves in order to form the full medallions, permanent symbols of the first international human spaceflight.

ASTP medallions: The astronauts gave the cosmonauts silver medallions in commemoration of their participation in the Apollo-Soyuz Test Project.

Tree seeds: The astronauts and cosmonauts exchanged tree seeds from their respective countries, which each nation could plant as a living and growing monument to the first cooperative human spaceflight.

FAI certificate of docking: The two crews carried aboard four copies of the International Aeronautical Federation (in French, Fédération Aéronautique Internationale; FAI) certificate of docking. The FAI has been the certifying agency for world air and space records since 1905. The certifi-

cate officially recorded the first docking between spacecraft from two nations.

"Agreement Concerning Cooperation in the Exploration and Use of Outer Space for Peaceful Purposes": The two crews carried aboard six copies of the 1972 agreement signed by U.S. president Richard M. Nixon and Soviet premier Alexei Kosygin, in which both nations made a commitment to conduct the Apollo-Soyuz Test Project and a wide range of continuing cooperative activities.

Lead-gold alloy: The two crews each brought aboard pieces of gold and lead, which they then melted and mixed in an electric furnace aboard the docking module. The unifor y (mixture of two unlike metals) was a new substance that symbolized the success of the nations in putting aside their differences to work together in space.

An almost tragic ending

Although the ASTP mission had realistically come to an end, the Apollo astronauts remained in space for another three days to conduct Earth observation studies and other experiments. On July 24, the astronauts donned their spacesuits, jettisoned the DM, and then prepared the craft for reentry. While the Apollo capsule was speeding through Earth's atmosphere, Brand forgot to operate the two switches that would automatically release the parachutes and shut down the thrusters. When the initial chute failed to come out, Brand was forced to hit the switch that opened it manually, causing the spacecraft to swing. The thrusters fired to correct the craft's swinging motion. Stafford noticed this and shut down the thrusters manually. However, during the thirty seconds that the thrusters were on, a mixture of toxic propellants from the thrusters had entered the cabin through a pressure relief valve.

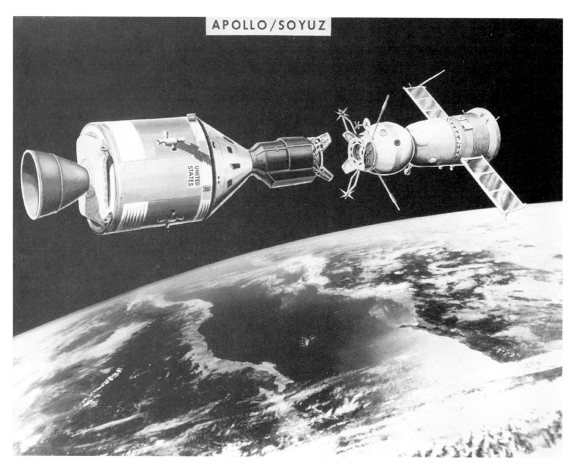

APOLLO/SOYUZ

Illustration of the midair docking of Apollo spacecraft (left) with a Soyuz spacecraft, high above Earth. *(National Aeronautics and Space Administration)*

Despite the choking gas fumes in the cabin, Brand was able to deploy the main parachutes, and the craft splashed down hard into the Pacific Ocean 4.5 miles (7.3 kilometers) from its recovery ship, turning upside down in the water. Brand was unconscious, and Stafford and Slayton nearly so. Stafford managed to place an oxygen mask over Brand's face, and he began to regain consciousness. Once the capsule righted itself, Stafford opened a valve to allow outside air into the cabin, and the remaining fumes disappeared. After they were rescued, the astronauts remained in a hospital for two weeks. Doctors discovered that the fast-acting gas had actually blistered their lungs. They also discovered a small lesion, or diseased area, on Slayton's left lung. After it was surgically removed, doctors determined that it was not cancerous. However, if the lesion had been discovered before the start of the mission, Slayton probably would have been denied his time in space.

Although the Apollo-Soyuz Test Project ended up a one-time-only event, it created a sense of goodwill for a time that exceeded the simple "handshake in space" that was its most visible symbol. Aside from its political significance, the mission resulted in a number of technical developments, including a common docking system that would be used in various forms on future missions into space. NASA considered a second ASTP mission in 1977, but worried that it would interfere with the developing space shuttle program. The "Agreement Concerning Cooperation in the Exploration and Use of Outer Space for Peaceful Purposes" was renewed in 1977, but the spirit of détente that had made ASTP possible evaporated following the Soviet invasion of Afghanistan in 1979.

ASTP Apollo was the last U.S. manned spaceflight that used a traditional rocket booster. It was also the last U.S. manned spaceflight until the first space shuttle launched in 1981.

For More Information

Books

Ezell, Edward Clinton, and Linda Neuman Ezell. *The Partnership: A History of the Apollo-Soyuz Test Project.* Washington, DC: National Aeronautics and Space Administration, 1978.

Froehlich, Walter. *Apollo Soyuz.* Washington, DC: National Aeronautics and Space Administration, 1976.

Slayton, Donald K., with Michael Cassutt. *Deke! An Autobiography.* New York: St. Martin's Press, 1995.

Web Sites

"Apollo-Soyuz: A Giant Leap in Cooperation." *CNN Interactive.* http://www.cnn.com/2000/TECH/space/07/17/apollo.soyuz/index.html (accessed on August 19, 2004).

"Apollo-Soyuz Test Project." *National Aeronautics and Space Administration History Office.* http://www.hq.nasa.gov/office/pao/History/astp/index.html (accessed on August 19, 2004).

"The Apollo Soyuz Test Project." *NASA/Kennedy Space Center.* http://www-pao.ksc.nasa.gov/kscpao/history/astp/astp.html (accessed on August 19, 2004).

"The Apollo-Soyuz Test Project." *U.S. Centennial of Flight Commission.* http://www.centennialofflight.gov/essay/SPACEFLIGHT/ASTP/SP24.htm (accessed on August 19, 2004).

"Apollo-Soyuz Test Project: Joint Mission in Space." *Smithsonian National Air and Space Museum.* http://www.nasm.si.edu/exhibitions/gal114/SpaceRace/sec500/sec520.htm (accessed on August 19, 2004).

"The Partnership: A History of the Apollo-Soyuz Test Project." *National Aeronautics and Space Administration History Office.* http://www.hq.nasa.gov/office/pao/History/SP-4209/toc.htm (accessed on August 19, 2004).

"Project Apollo-Soyuz Drawings and Technical Diagrams." *National Aeronautics and Space Administration History Office.* http://www.hq.nasa.gov/office/pao/History/diagrams/astp/apol_soyuz.htm (accessed on August 19, 2004).

10

Space Stations

Less than two decades after the first artificial satellite, *Sputnik 1,* lifted off into orbit on October 4, 1957, U.S. astronauts and Soviet cosmonauts were living and working in space stations hundred of miles above Earth's surface. The scientific foundations of such orbital outposts had been laid down almost seventy years previous; the dream of living in space had been imagined more than a century before.

A space station is a large orbiting structure designed for long-term human habitation in space. The first space station was a diversion, of sorts, from the space race, the contest to achieve superiority in spaceflight between the democratic, capitalist United States and the Communist Soviet Union (present-day Russia). By the end of the 1960s, after the United States had placed astronauts on the Moon, the Soviet Union quietly gave up its quest of a manned lunar (moon) landing and shifted its focus instead to the launching of the first space station.

In the last two-and-one-half decades of the twentieth century, a number of space stations were placed into Earth orbit, mostly by the Soviets. Although originally envisioned as a way station for piloted missions to the Moon and beyond (where

U.S. space station *Skylab* was placed into orbit on May 14, 1973.

(National Aeronautics and Space Administration)

they could stop to refuel or take on additional cargo), these vessels far surpassed that goal. They were, for the most part, orbiting laboratories in which groups of men and women carried out important scientific experiments and acted as subjects themselves in tests of the long-term effects of space on the human body.

American writer Edward Everett Hale (1822–1909), who was also an ordained Unitarian minister, is regarded as the first person to have put the idea of a space station into print. During his writing career, he wrote hundreds of pieces, including sermons, biographies, novels, and essays. Even within such forms as the short story, Hale produced a wide range of fiction, in various settings, for both children and adults. In the development of the genre of science fiction, he is considered a key figure, primarily on the strength of his story "The Brick Moon," which was originally published in 1869 in the *Atlantic Monthly*.

In the story, the character Frederic Ingham assembles a crew to build a 200-foot (61-meter) diameter sphere out of bricks that will serve as a fixed landmark for navigators on the seas. The sphere was to be launched into an orbit 4,000 miles (6,436 kilometers) high. At that height, navigators would see it from the surface of the planet as a bright star, much like the North Star. However, an accident causes the premature launching of the sphere and thirty-seven workers are thrown into space inside of the artificial moon. Once in orbit, they soon adapt well to their new life.

The fantastic idea of a space station was given a scientific grounding in the work of Russian scientist and rocket expert Konstantin E. Tsiolkovsky (1857–1935). Often referred to as the "father of astronautics," Tsiolkovsky had outlined by the beginning of the twentieth century many of the basic concepts and mathematical formulas of space travel that scientists still use in the present day. Like Hale, Tsiolkovsky used fiction to introduce his vision of a space station, but his story was simply a means to depict for a general audience what space travel and living in space would be like. In his 1920 science-fiction novel, *Beyond the Planet Earth,* Tsiolkovsky described for the first time a true space station, though he did not call it such. With living quarters for an international crew of six, the station featured a laboratory, a greenhouse, and a docking port for spacecraft.

Words to Know

Artificial satellite: A man-made device that orbits Earth and other celestial bodies and that follows the same gravitational laws that govern the orbit of a natural satellite.

Microgravity: A state where gravity is reduced to almost negligible levels, such as during spaceflight; commonly called weightlessness.

Solar flare: Temporary bright spot that explodes on the Sun's surface, releasing an incredible amount of energy.

Space motion sickness: A condition similar to ordinary travel sickness, with symptoms including loss of appetite, nausea, vomiting, gastrointestinal disturbances, and fatigue. The precise cause of the condition is not fully understood, though most scientists agree that the problem originates in the balance organs of the inner ear.

Space station: A large orbiting structure designed for long-term human habitation in space.

Spacewalk: Technically known as an EVA, or extravehicular activity, an excursion outside a spacecraft or space station by an astronaut or cosmonaut wearing only a pressurized spacesuit and, possibly, some sort of maneuvering device.

Sunspot: A cool area of magnetic disturbance that forms a dark blemish on the surface of the Sun.

The term "space station" was coined by another space visionary, German physicist Hermann Oberth (1894–1989). Like Tsiolkovsky, Oberth was a theorist. Although his practical experiments in rocketry were few, he helped popularize the concept of spaceflight as reality. In 1923 he published his recently rejected doctoral dissertation (a lengthy written statement on particular subject) as a nintey-two-page pamphlet titled *Die Rakete zu den Planetenräumen* (*The Rocket into Interplanetary Space*). The work was filled with calculations and even a design for a 115-foot-tall (35-meter-tall) bullet-shaped rocket with four large fins. Oberth also put forth the idea of a human expedition to Mars with an orbiting refueling station to be used as a staging point, or an area in which participants in a new mission are gathered and readied, for the voyage.

Six years later, Slovenian rocket engineer Herman Potocnik (1892–1929) published *Das Problem der Befahrung des Weltraums: Der Raketen-motor* (*The Problem of Space Travel: The Rocket Motor*). In the work, written under the pseudonym Hermann Noordung, he gave a detailed description of an orbiting space station shaped like a giant wheel that slowly rotated to produce artificial gravity. His nearly 100-foot-diameter (30-meter-diameter) station used solar mirrors to power a steam turbine that generated electrical power. The design included a donut-shaped structure for living quarters and an astronomical observation station. Potocnik addressed the problems of weightlessness, space communications, maintaining a livable environment for the crew, and extravehicular activities (EVAs) or spacewalks.

Potocnik's idea was expanded upon in the 1950s by German-born American engineer Wernher von Braun (1912–1977), who worked for the U.S. Army at the Redstone Arsenal near Huntsville, Alabama. Not only was von Braun a strong advocate of space exploration, he was a charismatic figure who drew people enthusiastically to his ideas and vision. Beginning in 1952 he worked with the popular *Collier's* magazine on a series of articles on space projects such as large rockets, lunar missions, and an orbiting space station. He then served as technical advisor on three television shows produced by Walt Disney Studios that, among other scientific and mechanical aspects of space travel, featured images of a wheel-like space station that served many purposes: as a laboratory, as an Earth-observation post, and as the launching point for missions to the Moon and Mars.

Von Braun's vision was soon adopted by officials from the National Aeronautics and Space Administration (NASA), who thought that a space station would be the best support for a wide range of space activities. The bold call by U.S. president John F. Kennedy (1917–1963) in 1961 to place a U.S. astronaut on the Moon and return him safely to Earth before the end of the 1960s, however, interrupted that plan. At the time, the race to beat the Soviets to the Moon was far more important than establishing a permanent presence in space.

NASA had hoped to return to its focus on a space station, and in 1969, the same year *Apollo 11* astronauts walked on the Moon, it did. Plans had already been drawn up for a

On April 19, 1971, the Soviets launched *Salyut 1,* **the first human-made space station.** *(© Bettmann/Corbis)*

proposed one-hundred-person space station that was to be completed by 1975. It was to have served as a laboratory for scientific experiments and as a base for nuclear-powered space-craft designed to carry people and supplies to and from an outpost on the Moon. NASA officials quickly realized, however, that the cost of building and supplying the station using conventional rockets would be far too expensive. What was needed were reusable spacecraft that could be used over and over again to ferry people and supplies to and from the station. And so NASA began work on those spacecraft, the space shuttles.

Salyut: The first space stations

Meanwhile, space engineers in the Soviet Union were transforming Tsiolkovsky's dreams into reality. On April 19, 1971, the Soviets launched *Salyut 1,* the first human-made space station. The Salyut (Russian for "Salute") program was a series of seven space stations launched over a period of eleven years. The purpose of the program was to make human presence in space routine and continuous. The stations could not be resupplied, so they had limited lifetimes in orbit. Astronauts from a variety of countries flew to the orbiting stations aboard Soyuz spacecraft.

Salyut 1 was a small station that could accommodate three cosmonauts for three to four weeks. It was shaped like a tube that was thinner in some parts than others, measuring 47 feet (14 meters) in length and 13 feet (4 meters) in diameter at its widest point. It weighed more than 25 tons (23 metric tons). Providing the station with power were four solar panels, which extended from its body like propellers. The station contained a work compartment and control center, a propulsion system, sanitation facilities, and a room for experiments. Cosmonauts entered the station through a dock at one end.

Three days after the launch of the station, *Soyuz 10* lifted off on the first mission to *Salyut 1*. However, the three cosmonauts aboard *Soyuz 10* could not successfully dock their spacecraft with the station and were forced to return to Earth. A second attempt to dock was then made by *Soyuz 11,* which carried cosmonauts Georgi Dobrovolski (1928–1971), Vladislav Volkov (1935–1971), and Viktor Patsayev (1933–1971) into space on June 6, 1971. The three cosmonauts successfully docked, then spent twenty-four days testing the space station's systems and conducting experiments. Their mission, however, ended in tragedy. During reentry, a pressure valve on the spacecraft remained open, allowing the air inside the capsule to escape. The crew suffocated to death. The Soviets then abandoned *Salyut 1* on October 11, 1971, allowing it to reenter Earth's atmosphere where it burned up.

The six subsequent Salyut space stations met with mixed success. Three of these—*Salyut 2, Salyut 3*, and *Salyut 5*—were military missions. These Salyuts were similar in design to the civilian Salyuts, but somewhat smaller, with only two solar panels. They also contained an unmanned capsule that could return to Earth material such as film or other sensitive information. All cosmonauts on missions to these stations were military pilots.

Salyut 2, the cover name of the highly secret Almaz military space station, was placed in orbit on April 3, 1973. The station quickly ran into trouble. Two days after launch, its flight control system failed and it lost pressure, making it uninhabitable for humans. The cause of the massive failure was likely due to metal fragments piercing the station when the discarded Proton rocket that had placed it in orbit later exploded nearby. No attempt was made to send a crew to the station, which eventually burned up in the planet's atmosphere.

A second Almaz, under the name *Salyut 3*, was launched on June 25, 1974. Cosmonauts Pavel Popovich (1930–) and Yuri Artyukhin (1930–1998) reached the station aboard *Soyuz 14*, docking with it on July 3. They spent two weeks on the station, conducting military tasks and various biomedical experiments. The follow-up crew lifted off aboard *Soyuz 15* on August 26. A disaster nearly occurred when a failure of the spacecraft's rendezvous system caused it to approach the station at the frightening speed of 45 miles (72 kilometers) per hour. Luckily, the spacecraft was off target and the crew was able to abort the mission, returning safely to Earth. The day before *Salyut 3* reentered the atmosphere on January 25, 1975, trials of an onboard aircraft cannon were conducted. A target satellite was destroyed in the successful test.

Salyut 4, a nonmilitary space station, lifted off on December 26, 1974. Two weeks later, the crew of *Soyuz 17*, cosmonauts Georgi Grechko (1931–) and Aleksei Gubarev (1931–), entered the station for a month-long stay. They conducted mostly astronomical experiments, making observations of the Sun, Earth, and the planets. The second crew destined for the

Russian cosmonauts Vladimir Aksenov and Yuri Malyshev relax before their mission to *Soyuz 6.* *(© Bettmann/Corbis)*

station never made it. Their spacecraft malfunctioned before they reached orbit, and they were forced to make an emergency landing in Siberia. A new crew aboard *Soyuz 18* lifted off for the station on May 24, 1975. Cosmonauts Pyotr Klimuk (1942–) and Vitali Sevastyanov (1935–) stayed on board *Salyut 4* for sixty-three days, setting a new space endurance record. In November of that year, an unmanned Soyuz spacecraft docked automatically with the station, remaining attached for three months. This demonstrated that supply missions to future space stations could be successful. *Salyut 4* reentered Earth's atmosphere on February 3, 1977.

The final Almaz military space station, *Salyut 5*, was launched on June 22, 1976. Three crews lifted off to serve aboard the station, but only two, those on *Soyuz 21* and *Soyuz 24*, successfully docked and boarded for lengthy stays. The crew launched aboard *Soyuz 23* never docked because of a failure of their craft's rendezvous system. Little is known of the activities of the two crews that served aboard *Salyut 5*. It is known that cosmonauts Boris Volynov (1934–) and Vitali Zholobov (1937–), who flew on *Soyuz 21*, had to cut short their stay on the station because they suffered from physical and psychological problems. Zholobov, in particular, was affected by intense space motion sickness and homesickness. *Salyut 5* reentered Earth's atmosphere on August 8, 1977.

Salyut 6, which launched on September 29, 1977, marked a turning point in space station design and technology. Although it resembled previous Salyut stations in overall design, it featured a second docking port in the rear, which allowed two spacecraft to dock with the station at the same time. An unmanned cargo spacecraft, known as *Progress,* could also dock at the second port, delivering fuel to the station's propellent tanks. Between 1977 and 1982, *Salyut 6* hosted five long-duration crews and eleven short-term crews, including cosmonauts from countries that were politically allied with the Soviet Union. Czech cosmonaut Vladimir Remek (1948–), who flew to the station aboard *Soyuz 28* (the third mission to dock with the station), was the first person launched into space who was not a citizen of either the United States or the Soviet Union. The very first long-duration crew on the station stayed ninety-six days in orbit. The longest stay on *Salyut 6,* made in 1980 by *Soyuz 35* cosmonauts Leonid Popov (1945–) and Valeri Ryumin (1939–), lasted 185 days. The five long-duration crews occupied the station for a total of 671 days.

When *Salyut 6* finally reentered Earth's atmosphere on July 29, 1982, *Salyut 7* had already been in orbit for three months. Launched on April 19, 1982, it stayed aloft for four years and two months, playing host to ten crews that consisted of six resident crews and four visiting ones (including French and Indian cosmonauts). A total of twenty-two cosmonauts visited the station, five of them twice and one three times. At any one time, two to six cosmonauts lived aboard *Salyut 7*. Cosmonauts Leonid Kizim (1941–), Vladimir Solovyov (1946–), and Oleg Atkov (1949–) spent 237 days on the station in 1984, the longest stay by any crew. While on board, they conducted experiments in astronomy and space manufacturing. Atkov, the first medical doctor to spend more than two days in space, also studied the effects of space travel on the human body. Solovyov made six EVAs during the mission to repair the station. *Salyut 7* burned up in the planet's atmosphere on February 7, 1991.

A U.S. experiment

Skylab was the only space station ever operated solely by the United States. It had sprung from the desire of NASA for a program that could apply hardware developed for Apollo lunar missions to other manned spaceflight objectives. The space agency thus approved an experimental manned Earth-orbiting laboratory as the program to follow Apollo. NASA officials hoped it would be the forerunner of a real space station. The laboratory was to be created inside the third stage of a Saturn V, the large rocket used to send Apollo spacecraft into orbit. Two laboratories were built, but before the first was ever launched, it was evident that cuts to NASA's budget due to the high cost of the Vietnam War (1954–75) and social programs would prevent the second from ever going into space. The total cost of the space station program was less than three billion dollars at the time. (In comparison, the total cost for the Apollo program was about twenty-five billion dollars; the Vietnam War cost the country more than 110 billion dollars and more than 58,000 lives.)

Skylab was placed into orbit on May 14, 1973, by the Saturn V, the last time that giant launcher was used. In orbit, the station was 118 feet (36 meters) long and weighed nearly 100 tons (91 metric tons) when an Apollo spacecraft was docked to it. The livable area of the station, known as the Orbital Workshop (OWS), was a cylinder 48 feet (17 meters) in length and 22 feet (6.7 meters) in diameter. Its volume, slightly more than 9,993 cubic feet (283 cubic meters), was about the same as that of a small house. It was divided into two levels separated by an open metal floor or grid into which the astronauts' cleated shoes could be locked. The upper floor had storage lockers and a large empty volume for conducting experiments, plus two scientific airlocks, one pointing down at Earth, the other toward the Sun. The lower floor had compartmented "rooms" with conditions that were far more comfortable than those of Apollo spacecraft: a dining room table placed beside a large window, a kitchen area with a freezer containing seventy-two different food selections and an oven of sorts, three bedrooms, a work area, and a shower and private bathroom custom-designed for use in the microgravity (very little gravity; near weightlessness) aboard the station. For example, the toilet had a seat belt to prevent the user from floating off.

Exercise equipment, including a stationary bicycle, was also provided to help the astronauts combat the loss of muscle tone caused by an extended stay in space. The only drawback to exercising in the station was that sweat floated off the astronauts' bodies in slimy puddles. Astronauts had to catch these puddles with a towel before they landed on a control panel or other piece of equipment, possibly causing harm.

The largest piece of scientific equipment, attached to one end of the Orbital Workshop, was the Apollo Telescope Mount (ATM). The solar observatory was used to study the Sun with no interference from Earth's atmosphere. The ATM had its own electricity-generating solar panels. *Skylab* also had an airlock module for EVAs that were required for repairs, experiment deployments, and routine changing of film in the ATM. The Apollo spacecraft that brought the astronauts to the station remained attached to the station's multiple docking adapter while the astronauts were on board.

The mission goals of *Skylab* were: to prove that humans could live and work in space for extended periods, and to expand knowledge of solar astronomy well beyond Earth-based observations. Because *Skylab* was a research laboratory, astronauts who served on missions to the station were different from those who had served aboard Mercury, Gemini, and Apollo missions. All previous astronauts had been pilots, except for one scientist on the last Apollo mission. The crews aboard *Skylab* included a number of scientist-astronauts.

Problems from the start

Almost immediately, *Skylab* encountered problems. Sixty-three seconds after liftoff, vibrations during the launch caused the meteoroid shield, which was designed to shade *Skylab*'s

U.S. astronaut Charles "Pete" Conrad Jr., commander of the first manned *Skylab* mission, during a training exercise in 1973.

(National Aeronautics and Space Administration)

OWS from the Sun's rays, to rip off. When it ripped off, it took with it one of the spacecraft's two solar panels. In addition, debris from the meteoroid shield wrapped around the other solar panel, keeping it from deploying properly.

Despite these problems, the station was able to achieve a nearly circular orbit around Earth, 270 miles (434 kilometers) above its surface. To provide energy to the station, NASA's mission control personnel maneuvered the solar panels of the ATM to face the Sun. Without the protective meteoroid shield, however, temperatures inside the OWS rose to 126°F (52°C).

The first crew (their mission was labeled *Skylab 2*) to occupy the station was to have launched the next day, but crew members waited on the ground for ten days while NASA engineers developed procedures and trained the crew to fix the crippled space station. Finally, on May 25, 1973, astronauts Charles "Pete" Conrad Jr. (1930-1999), Paul J. Weitz (1932–), and Joseph P. Kerwin (1932–) lifted off in an Apollo capsule atop a Saturn IB rocket (smaller than the Saturn V) and rendezvoused with the station. Their first priority was to lower the temperature inside to a comfortable level. After a failed attempt to deploy the stuck solar panel, they entered the station, thrusting a sunshade through an air lock to replace the lost thermal shield. The fix worked. Two weeks later, Conrad and Kerwin conducted a three-and-one-half hour EVA and, after a struggle, were able to free the stuck solar panel and restore power to the station. For nearly a month, they made further repairs to the OWS, conducted medical experiments, gathered solar and Earth science data, and returned some 29,000 frames of film. The astronauts spent twenty-eight days in space, doubling the previous U.S. record.

The second crew, *Skylab 3,* arrived at the station on July 28, 1973. However, all three astronauts—Alan L. Bean (1932–), Jack R. Lousma (1936–), and Owen K. Garriott (1930–)—fell victim to space motion sickness shortly after entering *Skylab.* Once the bout passed, they settled down to a fifty-nine-day stay on board the space station. During their mission, Garriott and Lousma deployed a second sun shield on an EVA lasting six-and-one-half hours. It was the first and longest of three EVAs the crew would make. During their two months in orbit, the astronauts undertook a busy schedule of experiments, including a student experiment to see if spiders could spin webs in weightlessness (they could). They also tested a jet-powered Astronaut Maneuvering Unit (AMU) backpack inside the spacious volume of *Skylab*'s forward compartment. The AMU, which had been carried but never flown on Gemini missions in the 1960s, proved a capable form of one-man space transportation.

An endurance record for the longest U.S. spaceflight was set by the crew of *Skylab 4:* eighty-four days, one hour, fifteen minutes, and thirty-one seconds. Following their November 16, 1973, launch, astronauts Gerald P. Carr (1932–), Edward G. Gibson (1936–), and William R. Pogue (1930–) carried out numerous experiments and set space records. They studied the effects of weightlessness; conducted biomedical experiments; observed and studied Earth, Comet Kohoutek, and a giant solar flare; and greatly increased humankind's knowledge of the Sun and its effect on Earth's environment. The astronauts also took four EVAs totaling twenty-two hours and twenty-two minutes, including one on Christmas Day to view Kohoutek. The longest of their EVAs lasted just more than seven hours, the longest spacewalk up to that time.

To help keep them in shape during their mission, a treadmill was added to the stationary bicycle already on board. As a result, the *Skylab 4* crew returned to Earth in better physical condition than previous crews. But an excessive work pace caused some tension during the mission. The astronauts conducted a total of 1,563 hours of scientific experiments. NASA flight controllers learned not to make excessive demands on the crew. At one point, the astronauts briefly refused to carry out their duties until a new schedule was negotiated with mission control.

Before leaving *Skylab,* the station's final crew boosted it to a slightly higher orbit, which varied from 267 to 283 miles (430 to 455 kilometers). NASA engineers calculated that this new altitude would allow *Skylab* to remain in orbit for at least nine more years.

The sad ending of a successful program

Despite its early mechanical difficulties, *Skylab* was an overwhelming success. Its three crews occupied the Orbital Workshop for a total of 171 days. They conducted nearly three hundred scientific and technical experiments, including medical experiments on how humans adapt to microgravity, solar observations, and detailed studies of Earth's resources. Both the time in space and the time spent on EVAs exceeded the combined totals of the entire world's previous spaceflights up to that time. Additionally, the ability to conduct longer missions was conclusively demonstrated by *Skylab,* as evidenced by the good health and physical condition of the second and third crews.

The end of the third mission to *Skylab* marked the end of the first phase of the program. NASA officials hoped the station would remain in orbit and would be reoccupied when the space shuttle program was under way. In the fall of 1977, however, NASA engineers determined that the station was no longer in a stable orbit as a result of greater-than-predicted solar activity. A space shuttle mission was planned for February 1980 in which astronauts would attach an upper stage to the station, boosting it into a higher orbit. On July 11, 1979, a year before the planned mission and two years before the shuttle's first actual flight, *Skylab* fell into the atmosphere and burned up over the Indian Ocean. Some debris from the station fell across the southeastern Indian Ocean and a sparsely populated section of western Australia. Luckily, no one was injured.

Skylab's flaming plunge to Earth marked the end of the Apollo era of human spaceflight.

Mir

The longest continuous presence of humans in space began in 1986 with the Soviet launch of a 20.9-ton (19-metric-ton) cylinder that formed the core of the space station called *Mir* (pronounced meer; Russian for "peace" or "community living in harmony"). By 1996 a total of six modules had been linked to complete the sprawling station. To build the space station, the Soviets (and their Russian successors) drew from lessons learned with the Salyut stations of the 1970s and 1980s. Those stations were simple and robust, but compact and with limited lifespans. Engineered from the beginning for expansion, *Mir* was designed to be resupplied regularly.

The longest continuous presence of humans in space began in 1986 with the Soviet launch of the space station called *Mir.* *(Digital image*
© 1996 Corbis; Original image courtesy of NASA/Corbis)

The heart of the station was the core module, placed in orbit on February 20, 1986. The core module had six ports for the attachment of other modules. These ports were placed in key locations, allowing the station's configuration to be changed. With attached solar panels generating power, the core module provided basic life support and command services. The 43-foot-long (13-meter-long) module consisted of a 10-foot-diameter (3-meter-diameter) cylinder attached to a 14-foot-diameter (4.2-meter-diameter) cylinder by a tapered segment. One side of the module housed living quarters; the other contained the space station operations, communications, and command center. The crew inhabiting the station (six at a time for short stays and three comfortably for longer periods) spent most of its time in the living and work areas. The living space consisted of two small sleeping cabins and a common area with dining facilities and exercise equipment. The space also contained a toilet, sink, and a water recycling system.

The next component connected to the station was the Kvant 1 module, launched into orbit on March 31, 1987. Divided into a pressurized laboratory compartment and a nonpressurized equipment compartment, the 19-foot-long (5.8-meter-long), 13.8-foot-diameter (4.2-meter-diameter) module was originally designated as the astrophysics research laboratory. It also contained a gyroscope-based assembly that operated off of solar energy to orient *Mir* in space without the use of precious fuel. At the far end of the module was a docking area for unmanned rocket drones that arrived from Earth at intervals to fit the station with supplies. Once emptied, then refilled with trash and waste, the drones were released to fall back toward the planet, burning up on reentry in the atmosphere.

Launched on November 26, 1989, the Kvant II module eventually docked with *Mir*'s core module on December 6. The 45-foot-long (13.7-meter-long), 14-foot-diameter (4.4-meter-diameter) module contained instruments and equipment such as an oxygen generation system and one that converted humidity in the *Mir* atmosphere into drinking water. In addition, it contained a toilet and shower facility. Through a complex series of filtration and processing steps, a unit on the module cleaned and recycled water from the sanitary facilities for reuse. Kvant II also featured a compartment with an airlock that allowed crew members to exit the complex for spacewalks.

The third addition to the *Mir* core module was launched on May 31, 1990. The 45-foot-long (13.7-meter-long), 14-foot-

U.S. Astronauts on *Mir*

Astronaut	Period	Total days
Norman E. Thagard	March 14, 1995–July 7, 1995	118
Shannon Lucid	March 22, 1996–September 26, 1996	188
John Blaha	September 16, 1996–January 22, 1997	128
Jerry Linenger	January 12, 1997–May 24, 1997	132
C. Michael Foale	May 15, 1997–October 5, 1997	145
David Wolf	September 25, 1997–January 31, 1998	128
Andrew Thomas	January 22, 1998–June 12, 1998	141

diameter (4.4-meter-diameter) Kristall module contained instruments used to produce high-technology equipment in the microgravity environment. It also housed a greenhouse designed to allow botanists to study the effects of microgravity conditions on plant growth. When the space shuttle began operations with *Mir* in 1995, the docking port that allowed the ship to mate with the station was attached at the far end of the Kristall module.

On May 20, 1995, the Spektr module flew into orbit and docked at the port opposite Kvant II. More than 39 feet (11.9 meters) long and 14 feet (4.4 meters) wide, the module was designed for surface studies of Earth and atmospheric research. It also provided living quarters for visiting astronauts from the United States and European countries. The module produced significant amounts of power for *Mir* from four 370-square-foot (34.4-square-meter) solar panels.

The final unit segment of *Mir* was the Priroda module, placed in orbit on April 23, 1996. That was ten years after the core module had been placed in space and five years beyond the planned lifetime of the station. The module, similar in shape and size to the other modules, housed radar systems and detectors to study Earth's oceans and atmosphere. Lacking solar panels, Priroda was unable to generate its own power, relying instead on batteries or on the power network of *Mir*.

By the end of construction, *Mir* weighed 135 tons (122 metric tons) and offered 9,993 cubic feet (283 cubic meters) of space. This meant that, with the exception of the Moon, *Mir* was the heaviest object in orbit around Earth. Over its lifetime, its maintenance cost continued to skyrocket, and the station ultimately cost 4.2 billion dollars to construct and maintain. The space station was neither designed nor constructed to last the fifteen years it spent orbiting Earth. It far surpassed records set by *Skylab* for time in space.

With the fall of the Soviet Union in 1991, *Mir* became more expensive than what the new Russian nation could afford. Over the next ten years, the station deteriorated with age and became more difficult to fix. It suffered from problems with its insulation and glitches during docking and undocking procedures with supply craft.

Collaboration on a troubled station

In 1994, the United States made an historic four-hundred-million-dollar deal with Russia to place U.S. astronauts on *Mir* for durations of up to six months. NASA and the Russian space agency agreed to develop the future International Space Station (ISS). In preparation for that project, the two agreed to engage in a series of joint missions involving *Mir* and the space shuttle. Many heralded the Shuttle-Mir mission as the beginning of an era of continuing cooperation between the United States and Russia in space. Critics, however, argued that it was simply an underhanded funding of the troubled Russian space program. They further argued that it exposed U.S. astronauts to unnecessary and unacceptable risks.

Of the seven U.S. astronauts who eventually served on *Mir*, Shannon Lucid (1943–) stayed aboard the longest at 188 days. Fifty-three years old at the time, she set a new U.S. record for long-duration spaceflight. Lucid, who had become an astronaut in August 1979, had also served on four space shuttle flights previous to her time on *Mir,* making her the first woman to go into space more than twice. In all, she logged 223 days in space, the most by any woman.

Despite the wear and tear of more than a decade in space, *Mir* functioned surprisingly well until 1997. That year brought mishap after mishap. In February, an oxygen canister burst into flames, filling the living module with smoke. When crew

U.S. astronauts Linda Godwin (standing left), Kevin Chilton (standing center), Shannon Lucid (seated right), and Yuri Onufrienko aboard the Russian space station *Mir*, 1996. *(AP/Wide World Photos)*

members turned to extinguish the flames, they discovered that the launch restraints on the firefighting equipment had never been removed. Valuable time was spent searching in near darkness for tools to free the equipment before the flames were finally extinguished. A few weeks later, the main carbon dioxide removal system failed. Then the cooling system malfunctioned, leaking coolant into the air. Temperatures in the modules remained at 96°F (36°C) for weeks.

In June *Mir* suffered its most dangerous setback. During the testing of a new docking system, an unmanned rocket collided with the Spektr module, piercing the hull and crumpling solar panels. During the scramble to seal off the module, crew members were forced to disconnect cables that snaked from Spektr's solar panels into the other modules of the station,

Top Ten Single Flight Durations

Cosmonaut	Station	Period	Total days
Valeri Polyakov	*Mir*	January 8, 1994–March 22, 1995	437
Sergei Avdeyev	*Mir*	August 13, 1998–August 28, 1999	379
Musa Manarov	*Mir*	December 21, 1987–December 21, 1988	365
Vladimir Titov	*Mir*	December 21, 1987–December 21, 1988	365
Yuri Romanenko	*Mir*	February 6, 1987–December 29, 1987	326
Sergei Krikalyov	*Mir*	May 18, 1991–March 25, 1992	311
Valery Polyakov	*Mir*	August 29, 1988–April 27, 1989	240
Leonid Kizim	*Salyut 7*	February 8, 1984–October 2, 1984	237
Vladimir Solovyov	*Salyut 7*	February 8, 1984–October 2, 1984	237
Oleg Atkov	*Salyut 7*	February 8, 1984–October 2, 1984	237

leaving *Mir* with only partial power. Days later, the steering units broke down, then a power surge knocked out a computer. Crew members were forced to use precious fuel from the Soyuz escape pod to reposition the station, turning the solar panels toward the Sun.

The following month, the cooling system failed yet again. Then the main computer crashed, an event that would repeat itself again and again in coming months. In September U.S. mission control sent out a warning that a military satellite was in an orbit that would pass dangerously close to *Mir*. At about the same time the warning was received, the main computer on the station failed, leaving the crew members on board to watch tensely from the Soyuz escape pod as the satellite passed only 3,000 feet (914 meters) away.

As the month dragged on, the station suffered repeated computer failures, as well as the failure of the carbon dioxide removal system and leaks of mysterious brown fluid. Concern for the safety of U.S. astronauts aboard the station mounted. The four-year collaboration ended when Andrew Thomas

(1951–), the seventh and final U.S. astronaut to serve aboard *Mir,* flew back to Earth in June 1998.

With the launch of the International Space Station approaching and continued problems on *Mir,* the Russian space agency announced plans to de-orbit the station in September 1999. As the appointed date drew near, however, the space agency seemed less and less inclined to terminate the station. Offers came in from different groups to try to save it. One group of entrepreneurs tried to turn *Mir* into a destination for wealthy tourists. Millionaire Dennis Tito (1940–) offered to pay twenty million dollars to become the first tourist aboard *Mir,* but his offer was not accepted before Russia ultimately decided to end the fifteen-year saga of the station. By that time, the station's orbit was degrading by almost 1 mile (1.6 kilometers) per day. After much planning, the Russian space agency decided to send *Mir* through Earth's atmosphere, allowing it to break apart into small pieces before its final splashdown in the South Pacific.

On March 23, 2001, after more than 86,000 orbits around the planet, *Mir* entered the atmosphere, breaking up into several large pieces and thousands of smaller ones. The larger pieces splashed down safely into the ocean.

A true international space station

Mir had been the first permanently crewed space station designed as an assembly, or complex, of specialized research modules. The five modules had been added one at a time. Even while beginning the assembly and operation of *Mir,* the Soviets were planning another *Mir*-type station. It was a plan revised because of developments both at home and in the United States.

In his State of the Union address before a joint session of the U.S. Congress on January 25, 1984, President Ronald Reagan (1911–2004) directed NASA "to develop a permanently manned space station and to do it within a decade." He went on to say that "NASA will invite other countries to participate." So began the International Space Station (ISS) project and, indirectly, the coming together of Soviet and U.S. space station projects.

Artist's rendering of the International Space Station in its planned completion. *(© Stocktrek/Corbis)*

NASA had wanted to undertake a permanent space station project since the space shuttle flight program began in 1981. Preliminary design studies were already under way when the president made his announcement. Within weeks, NASA invited other countries (except the Soviet Union) to join the project. Interest was already high at the European Space Agency (ESA), a multinational organization composed of fifteen member states dedicated to the exploration of space. The space agencies of Canada and Japan were also interested in participating in what was then called Space Station Freedom.

When it collapsed and broke apart in 1991, the former Soviet Union (now called Russia) was eventually invited to join the effort. Since the Russian space agency faced severe financial problems (as did all of Russia after the breakup), it accepted help from the United States and eventually agreed to join and lend its vast experience to the creation of a truly international space station.

In 1993 the United States put forth a detailed long-range ISS plan that included substantial Russian participation as well as the involvement of fourteen other nations. Altogether sixteen countries—Belgium, Brazil, Canada, Denmark, France, Germany, Italy, Japan, the Netherlands, Norway, Russia, Spain, Sweden, Switzerland, the United Kingdom, and the United States—banded together on a nonmilitary effort so complex and expensive that no one nation could ever consider doing it alone. The project, involving more than one hundred thousand people in space agencies and contracting companies around the world, was expected to be completed by 2006. The estimated lifetime cost of the station was one hundred billion dollars.

The ISS project has lofty goals. It is expected that having long-term, uninterrupted access to outer space will allow investigators to acquire large sets of data in weeks that would have taken years to obtain. The ISS project also plans to conduct medical and industrial experiments that it hopes will result in benefits to all humankind.

Largest adventure into space
The ambitious ISS has been likened in difficulty to building a pyramid in the zero gravity of space. When completely assembled, the ISS will have a mass of nearly 1 million pounds (454,000 kilograms) and will measure about 360 feet (110 meters) across by 290 feet (88 meters) long, making it much wider than the length of a football field. This large scale means that it can provide 46,000 cubic feet (1,300 cubic meters) of pressurized living and working space for a maximum crew of seven scientists and engineers. This amount of usable space is equal in size to the volume of a huge Boeing 747 jumbo jet. This massive structure will get its power from nearly 26,880 square feet (2,500 square meters) of solar panels spread out on four modules. These panels always rotate to face the Sun and can convert sunlight into electricity that can be stored in batteries. The station will have fifty-two computers controlling its numerous systems.

Canadian mission specialist Chris A. Hadfield works near a robotics tool outside of the International Space Station. *(National Aeronautics and Space Administration)*

The main components of the ISS are the service module and six scientific laboratories: the U.S. Destiny laboratory, the U.S Centrifuge Accommodation Module, the ESA Columbus Orbital facility, the Japanese Experiment Module, and two Russian Research Modules. The other major contributor is Canada, which provided a 55-foot-long (16.7-meter-long) robotic arm for assembly and other maintenance tasks. The United States also has the responsibility for developing and ultimately operating all of the major elements and systems aboard the station. More than forty spaceflights will be required to deliver more than one hundred separate components to the station. As of 2004, flights had been made by the space shuttle and the Soyuz and Progress spacecraft. The ISS orbits Earth at an average altitude of 220 miles (354 kilometers). At this distance, it can make observations of 85 percent of the planet and fly over 95 percent of Earth's population.

On November 11, 1998, the Russians placed the first major piece of the puzzle, the control module named Zarya, in orbit. Following the launch of the U.S. module named Unity during December of that year (which would serve as the connecting passageway between sections), the Russians launched their service module named Zvezda on July 12, 2000. This not only provided life-support systems to other elements but also served as early living quarters for the first crew. After more flights to deliver supplies and equipment, the U.S. laboratory module named Destiny was docked with the station on February 7, 2001. This state-of-the-art facility will be the centerpiece of the station when it is complete. The aluminum lab is 28 feet (8.5 meters) long and 14 feet (4.3 meters) wide and allows astronauts to work in a comfortable climate all year round.

Research on the station

The main goal of the ISS project is to conduct long-term scientific research in space. Astronauts will test themselves and learn more about the effects of long-term exposure to reduced gravity on humans. Studying how muscles weaken and what changes occur in the heart, arteries, veins, and bones may not only lead to a better understanding of the body's systems, but also might help humans plan for future long-term exploration of the solar system.

Flames, fluids, and metals all act differently in microgravity, and astronauts will conduct research in what is called Materials Science to try to create better alloys (metals created by the mixing and fusing of two or more different metals). The nature of space itself will be studied by examining what happens to the exterior of a spacecraft over time. Lastly, Earth itself will be watched and examined. Studying its forests, oceans, and mountains from space may lead to a better understanding of the large-scale, long-term changes that take place in the environment, especially those caused by air and water pollution and by the cutting and burning of forests. The ISS will have four large windows designed just for looking at Earth.

Space Tourists

On April 30, 2001, U.S. investment broker Dennis Tito entered the International Space Station (ISS) after having flown aboard a Soyuz spacecraft. Four hundred and fourteen people had flown into space before him, including a Saudi prince, a Russian politician, a Japanese television reporter, and three U.S. congressmen. But at a cost of twenty million dollars, Tito was the first paying tourist.

NASA had been less than thrilled about sending a nonprofessional into space, fearing he would jeopardize work on the station. The agency sought to have Tito barred from the trip. But the Russian space agency, strapped for cash, gladly accepted his money. It did require Tito to complete nine hundred hours of training and medical tests at Star City, the cosmonaut training center in Russia, before he would be approved to fly. Tito also had to agree to replace any equipment he broke during his week-long stay aboard the station.

After Tito's trip, the world's five biggest space agencies established health and training standards for both astronauts and visitors to the ISS.

In April 2002, South African Internet entrepreneur Mark Shuttleworth became the second paying tourist to board the ISS. His twenty-million-dollar trip also made him the first African to travel into space. Like Tito's, his trip had been arranged with the Russian space agency through Space Adventures, the world's leading space experiences company. In March 2004, the company announced that U.S. technology entrepreneur Gregory Olsen would become the third space tourist. The launch date for his expedition to the ISS was planned for April 2005.

The first crew, consisting of U.S. astronaut William Shepherd (1949–) and Russian cosmonauts Yuri Gidzenko (1962–) and Sergei Krikalev (1958–), entered the ISS on November 2, 2000, and stayed aboard until March 14, 2001. Since then, the station has been permanently crewed, with each outgoing crew handing over the ISS to the incoming crew. As of mid-2004, nine missions crewed by twenty-four astronauts or cosmonauts have been sent to the ISS. More than eighty visitors have also occupied the ISS, including the world's first space tourists. This makes it the most visited spacecraft in the history of spaceflight.

But overruns and cutbacks have occurred, threatening the continuation of what has become the world's only remaining space station. The project has been far more expensive than NASA originally anticipated, and construction is behind schedule. Because of this, the station cannot accommodate its expected crew of seven. This has limited the amount of science that can be performed on the station. Critics have labeled the project a waste of time and money, which could have been better spent on problems on Earth. After the accident of the space shuttle *Columbia* on February 1, 2003, the U.S. space shuttle program was suspended. This has halted construction of the ISS, since the shuttles delivered almost all the equipment and materials. Although Soyuz spacecraft continue to exchange crews who monitor the station and make repairs, the future of the ISS remains uncertain.

For More Information

Books

Caprara, Giovanni. *Living in Space: From Science Fiction to the International Space Station.* Buffalo, NY: Firefly Books, 2000.

Harland, David M. *The MIR Space Station: A Precursor to Space Colonization.* New York: Wiley, 1997.

Harland, David M., and John E. Catchpole. *Creating the International Space Station.* New York: Springer Verlag, 2002.

Launius, Roger D. *Space Stations: Base Camps to the Stars.* Washington, DC: Smithsonian Institution Press, 2003.

Shayler, David J. *Skylab: America's Space Station.* New York: Springer Verlag, 2001.

Web Sites

"International Space Station." *Boeing.* http://www.boeing.com/defense-space/space/spacestation/flash.html (accessed on August 19, 2004).

"International Space Station." *National Aeronautics and Space Administration.* http://spaceflight.nasa.gov/station/ (accessed on August 19, 2004).

"Living and Working in Space." *NASA Spacelink.* http://spacelink.nasa.gov/NASA.Projects/Human.Exploration.and.Development.of.Space/Living.and.Working.In.Space/.index.html (accessed on August 19, 2004).

Mir. http://www.russianspaceweb.com/mir.html (accessed on August 19, 2004).

"Skylab." *NASA/Kennedy Space Center.* http://www-pao.ksc.nasa.gov/kscpao/history/skylab/skylab.htm (accessed on August 19, 2004).

11

Space Shuttles

Perhaps no U.S. space program has borne witness to such great triumph and tragedy as that of the space shuttle program. Known officially as the Space Transportation System (STS), the space shuttle program was first put into operation in 1981. Between that time and early 2003, 113 shuttle missions were flown, carrying a total of 660 crew members. Many of those missions were marked by U.S. space firsts: Shuttles carried aloft the first U.S. female astronaut, the first African American male and female astronauts, the first U.S. female mission commander and pilot, the first Hispanic astronaut, and the first Native American astronaut.

Before the space shuttle, launching cargo into space was a one-way proposition. Rockets were used to put a tiny capsule carrying human space travelers into orbit. Stage by stage, booster segments of the rocket would fall away during the launch as their fuel ran out. The spacecraft would go into orbit around Earth, and then it would fall back to Earth, plunging into the ocean. At that point, it became space rubbish. Every part of the vehicle was discarded, never to be used again, with the exception of the human crew. Satellites could also

The space shuttle *Columbia* **moving from the vehicle assembly building to the launch pad in 1994.** *(© Corbis)*

be sent into orbit the same way as astronauts were, but they could not return.

The space shuttle, the world's first reusable space vehicle, changed that. It revolutionized the way people worked in space. Astronauts aboard space shuttle missions not only released satellites into orbit, they also captured, repaired, and then redeployed those already in space, the first time that had ever been accomplished. On April 25, 1990, astronauts aboard the shuttle *Discovery* deployed the Hubble Space Telescope. Over the next twelve years, four more shuttle missions repaired and upgraded the orbiting telescope. Eleven space shuttle missions also docked with the *Mir* space station and sixteen with the International Space Station, carrying supplies and crew members to and from the stations.

Astronauts aboard space shuttles have carried out a wide variety of tasks. In addition to the launching of scientific, commercial, and military satellites, shuttle crews have launched interplanetary probes. They have also conducted research in areas such as astronomy, biology, and space medicine.

The space shuttle changed the social makeup of space travelers. With such large crews (sometimes up to seven members on a mission), shuttle astronauts were divided into three categories: commander and pilot, mission specialist, and payload specialist. The commander and the pilot are both pilot astronauts. The commander has responsibility for the entire shuttle, crew, mission, and, most important, flight safety. The pilot assists the commander in operating the vehicle. A mission specialist is an astronaut who works with the commander and pilot and is responsible for crew activity planning, experiments, and the operation of any payload (any cargo launched aboard a spacecraft, including astronauts, instruments, and equipment). Mission specialists also perform spacewalks, or EVAs. A payload specialist may not be a professional astronaut employed by the National Aeronautics and Space Administration (NASA). Nominated by NASA, a foreign government or agency, a U.S. government agency, or a commercial company, a payload specialist has particular onboard duties relating to a specific payload on the mission. That person must have the required education and physical skills necessary to complete the mission.

But with the giant leaps forward in manned spaceflight have come tragic setbacks. Over the course of the program, two shuttles have been lost: *Challenger* in 1986 and *Columbia* in 2003. In total fourteen astronauts perished. Both disasters left NASA reeling and searching for stability, respect, and direction. The subsequent investigations into the disasters scolded NASA for its emphasis on style and gimmickry over science and safety. Blame was also placed on NASA's management style. The organization was ill prepared to fend off such attacks that tarnished its image. After the *Challenger* explosion, another shuttle did not fly for more than two years. After *Columbia*'s doomed flight, it was expected that another shuttle would not fly until at least the spring of 2005.

The most complex machine

The space shuttle is the most complex machine ever built. It is composed of more than 2.5 million parts, including four main components: the orbiter, three main engines, an external fuel tank, and two solid rocket boosters. These are the parts of the vehicle seen when the shuttle is launched. The combined weight of the vehicle at launch is approximately 4.5 million pounds (2 million kilograms). To lift the entire vehicle into the air very quickly requires about 7.3 million pounds (32.5 million Newtons) of thrust (the forward force generated by a rocket). One pound (4.45 Newtons) of thrust is the amount of thrust it takes to keep a 1-pound (0.454-kilogram) object stationary against the force of gravity on Earth. (Newton is the official metric unit of measure of force, named for English physicist and mathematician Isaac Newton [1642–1727].)

The delta-winged orbiter is the main part of the space shuttle. Constructed of aluminum, it is similar in size to a DC–9

Words to Know

Interplanetary medium: The space between planets including forms of energy and dust and gas.

Microgravity: A state where gravity is reduced to almost negligible levels, such as during spaceflight; commonly called weightlessness.

Payload: Any cargo launched aboard a spacecraft, including astronauts, instruments, and equipment.

Probe: An unmanned spacecraft sent to explore the Moon, other celestial bodies, or outer space; some probes are programmed to return to Earth while others are not.

Propellant: The chemical mixture burned to produce thrust in rockets.

Solar wind: Electrically charged subatomic particles that flow out from the Sun.

Space shuttle: A reusable winged spacecraft that transports astronauts and equipment into space and back.

Thrust: The forward force generated by a rocket.

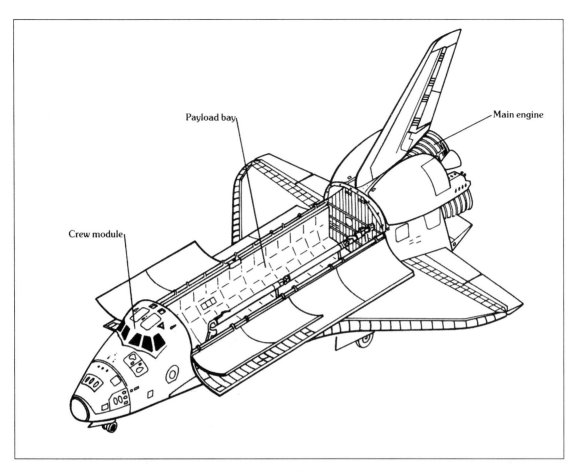

The delta-winged orbiter, the main part of the space shuttle, is divided into two parts: the crew cabin, or module; and the cargo, or payload, bay. *(National Aeronautics and Space Administration)*

commercial jet airliner. It has a length of 121 feet (37 meters), a wingspan of 78 feet (24 meters), and stands at a height of 56 feet (17 meters). The orbiter is divided into two parts: the crew cabin at its forward section and the cargo bay in its middle. The crew cabin contains the flight deck (sometimes called the flight control center) and living quarters for the crew. The forward portion of the flight deck resembles the cockpit of a jet airplane, but it features separate controls for flying both in space and through the planet's atmosphere. The crew quarters deck, lined with ten windows, has facilities for eating, sleeping, and sanitation. The orbiter is designed to carry from two to eight crewmembers on a ten-to-fourteen-day mission.

During launch, up to four astronauts may sit on the upper flight deck and up to four more may sit on the middle crew quarters deck.

The long cargo bay that takes up the majority of space on the orbiter contains the payload, which represents the purpose for the mission and "pays" for the flight. The cargo bay measures 60 feet (18.3 meters) in length, 15 feet (4.6 meters) in width, and 13 feet (4 meters) in depth. It can carry up to 65,000 pounds (29,510 kilograms) of material into space. The Hubble Space Telescope, for example, filled the entire orbiter cargo bay. Two large doors protect the cargo bay during ascent and descent. During operations in orbit, the doors must remain open to provide cooling for the orbiter. If the doors cannot open, the orbiter must return to Earth within eight hours. The cargo bay also contains a large robotic arm called the Remote Manipulator System (RMS). It measures 50.2 feet (15.3 meters) in length and 15 inches (38 centimeters) in diameter. The arm is articulated, meaning it has joints that allow it to move in a fashion similar to the human arm. From a control station at the rear of the flight deck, an astronaut operates the RMS to deploy satellites from the cargo bay or to retrieve and repair those already in orbit. The RMS can also function as an extension ladder for astronauts during EVAs.

Protecting the orbiter from the heat generated during reentry through Earth's atmosphere is vital. The survival of the astronauts onboard depends on it. This protection rests upon 32,000 silica-fiber thermal tiles that cover various areas of the orbiter. These square tiles vary from a measurement of 6 by 6 inches (15 by 15 centimeters) to 8 by 8 inches (20 by 20 centimeters). They range in thickness from 1 to 3 inches (2.5 to 7.6 centimeter) and have a consistency similar to chalk.

The areas most likely to encounter intense heat, such as the bottom of the orbiter and its nose, are covered by 20,000 tiles coated with black glass that can resist temperatures up to 2,300°F (1,260°C) by radiating 90 percent of the heat back into the atmosphere. Tiles that cover the upper side of wings and the sides closest to the nose are coated with white glass and can resist temperatures up to 1,200°F (650°C). The orbiter's nose and leading edges of the wings receive the most intense heat. These surfaces are covered tiles made from a material called reinforced carbon-carbon that can withstand temperatures up to 2,965°F (1,630°C). Areas that encounter only mild heat, including the top of the wings and the cargo bay, are covered with a thin layer of white insulation that provides protection up to 700°F (371°C).

The three main engines, mounted at the rear of the orbiter in a triangular pattern, are the most advanced liquid-propellant rocket engines ever manufactured. They provide the thrust necessary for the orbiter to achieve orbit. They each measure 14.1 feet (4.3 meters) in length and 7.5 feet (2.3 meters) in diameter and are designed to operate for 7.5 accumulated hours. These engines burn a propellant of liquid hydrogen and liquid oxygen. The liquid hydrogen is –423°F (–253°C), the second-coldest liquid on Earth. (Liquid helium, which exists only at temperatures near absolute zero, –459°F [–273°C], is the coldest liquid in nature.) When the liquid hydrogen is burned with the liquid oxygen, the temperature in the engine's combustion chamber reaches more than 6,000°F (3,316°C), about two-thirds the temperature of the surface of the Sun.

Each engine can provide up to 375,000 pounds (1,668,750 Newtons) of thrust at sea level. The engines may be throttled (to regulate the speed of an engine) from as low as 65 percent of their rated thrust to as high as 109 percent. A thrust value of 104 percent, known as full power, is typically used when the shuttle ascends, or climbs, through Earth's atmosphere. In an emergency, the engines may be throttled up to 109 percent. This throttling ability makes these engines much more efficient than all previous rocket engines, which could only deliver either 0 percent or 100 percent of their rated thrust.

The three engines are the most complicated and dangerous parts of the shuttle, which cannot attain orbit if the engines do not work perfectly. After operation, the engines are usually removed from the orbiter, inspected, tested, and then put into a rotation to be used in a future shuttle flight. Therefore, each orbiter is normally fitted with different engines prior to its next flight.

All of the fuel for the orbiter's ascent is contained in the external tank, the largest and only nonreusable element of the shuttle. Measuring 154 feet (47 meters) in length and 28 feet (8.5 meters) in diameter, the external tank is located between the two solid rocket boosters. The orbiter itself sits piggyback on the external tank. To keep ice from forming on the outside of the external tank (due to the very low temperature of the liquid fuel inside), it is covered with a thin layer of burnt-orange foam. In the external tank are two smaller tanks: a top tank containing 145,000 gallons (549,000 liters) of liquid oxygen and a bottom tank containing 388,000 gallons (1,470,000 liters) of liquid hydrogen. (Even though there is a greater volume of liquid hydrogen, it actually weighs one-quarter of the liquid oxygen because oxygen is sixteen times heavier than hydrogen.) The liquid oxygen and liquid hydrogen are supplied to the orbiter's three engines through 17-inch-diameter (43-centimeter-diameter) pipes. When filled with fuel, the external tank weighs approximately 1,655,600 pounds (751,640 kilograms); when empty, it weighs approximately 65,500 pounds (29,700 kilograms).

The solid rocket boosters, which contain the largest solid-propellant motors ever built and the first designed to be reused, measure 149 feet (45.5 meters) high and 12 feet (3.7 meters) in diameter. At liftoff, each solid rocket booster produces 3.3 million pounds (14.7 million Newtons) of thrust, just more than 70 percent of the thrust necessary to launch the space shuttle. Each solid rocket booster consists of four segments of solid propellant stacked vertically with a nose cone on top. The nose cone contains the propellant igniter, electronic devices that communicate with the orbiter, and parachutes that allow the boosters to be recovered at sea after they are released. The propellant is a mixture of 69.6 percent

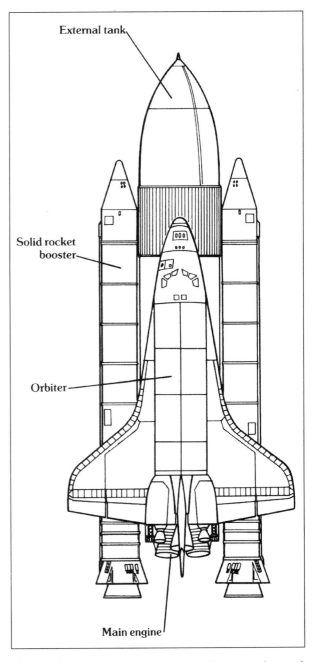

The shuttle's three main engines are the most advanced liquid-propellant rocket engines ever manufactured. All of the fuel is contained in the external tank. The solid boosters contain the largest solid-propellant motors ever built. *(National Aeronautics and Space Administration)*

ammonium perchlorate oxidizer, 16 percent aluminum powder fuel, 12 percent polymer binder, 2 percent epoxy curing agent, and 0.4 percent iron oxide to help control the burning rate. The mixture of these materials looks like a thick plaster. After it is poured into a mold and dried for several days, it looks and feels like a rubber eraser. When filled with fuel, each booster weighs approximately 1,300,000 pounds (590,200 kilograms); when empty, each one weighs approximately 192,000 pounds (87,170 kilograms).

Assembly and launch sequence

The space shuttle begins its journey into space from the launch area at the Kennedy Space Center at Cape Canaveral, Florida. Each space shuttle flight requires years of mission planning and months of preparing or processing the shuttle to go into orbit. Every system on the orbiter is inspected in a process that takes four to six weeks. Any items failing the rigorous exam are repaired or replaced. This includes each one of the shuttle's 32,000 thermal tiles.

At the end of the inspection process, the orbiter is lowered onto a mover and towed to a vehicle assembly building. At 525 feet (160 meters) high, it is the world's largest building in volume under a single roof. Here the orbiter is attached to the external tank, and the external tank is attached to the solid rocket boosters, which have all undergone their own inspection activities prior to their arrival in the vehicle assembly building.

Ready to be transported to one of the two shuttle launch pads, the completed vehicle is hoisted onto a flatbed vehicle called a crawler. It weighs 6 million pounds (2.7 million kilograms), is 131 feet (40 meters) long, 114 feet (35 meters) wide, and 10 feet (3 meters) tall. The crawler then moves along a 40-foot-wide (12-meter-wide) gravel road to the launch pad. The 3.4-mile (5.5-kilometer) journey takes 6 hours. At the launch pad, the crawler places the vehicle on a base of support posts next to the launch tower, then departs.

Countdown terminology is used to provide a rough guideline for going over the mandatory checklist items prior to launch. The term "T-minus" denotes the time remaining in the countdown. Liftoff occurs at T-minus zero seconds. The term "T-plus" denotes the time after liftoff.

On the day of launch, the shuttle and the attached external tank and solid rocket boosters sit vertically on the launch pad. At approximately T-minus thirty seconds, the shuttle's on-board computers take over control of the launch sequence. At T-minus 6.6 seconds, the shuttle's three main engines are ignited one at a time, 0.12 seconds apart. They quickly build to more than 90 percent of their rated thrust. At T-minus zero seconds, the liquid propellants from the external tank are pressure fed at a combined rate of 1,035 gallons (3,917 liters) per second through pipes to the orbiter's three main engines. The motors in the solid rocket boosters are also ignited (they cannot be shut down once they are ignited), burning 10,000 pounds (4,536 kilograms) of propellant per second.

The shuttle lifts off the pad, clearing the launch tower at T-plus three seconds. At T-plus twenty seconds, the shuttle begins its roll sequence so that it can enter the correct orbital path. It slowly begins to roll over to fly in a more easterly direction rather than merely straight upward. In this position, the orbiter's cargo bay faces Earth while the external tank is above it.

The space shuttle *Challenger* on the takeoff pad. Known officially as the Space Transportation System (STS), the space shuttle program was first put into operation in 1981. *(AP/Wide World Photos)*

At T-plus fifty seconds, the shuttle has reached an altitude of about 6.6 miles (10.6 kilometers). To protect the shuttle from aerodynamic stress caused by the atmospheric pressure of the air around it and from excessive heating, the shuttle's main engines are throttled back to 67 percent at this point. At about T-plus seventy seconds, the engines resume full throttle.

At approximately T-plus 1.42 minutes, at an altitude of about 28 miles (45 kilometers), the solid rocket boosters burn out and are jettisoned, or released. Milliseconds later, sixteen solid-fueled separation motors on the boosters are fired briefly to help carry the boosters away from the shuttle. About 3.75 minutes after separation, when the boosters are at an altitude of about 15,750 feet (4,800 meters), the nose cone of each booster is jettisoned and parachutes are deployed. The boosters then safely splash down in the Atlantic Ocean about 140 miles (225 kilometers) from the launch site where they are recovered by two NASA ships, *Liberty* and *Freedom*.

At T-plus 4.5 minutes, the shuttle can no longer return to the Kennedy Space Center if an engine problem or some other emergency develops. Over the next two minutes, the shuttle climbs from an altitude of 64 miles (103 kilometers) to 75 miles (121 kilometers). Also during this time the shuttle's trajectory, or curved path through the atmosphere, flattens so the tops of the astronauts' heads are pointing directly to the ground. By T-plus 6 minutes, the shuttle has a horizontal velocity of 12,400 miles (19,950 kilometers) per hour. This speed, however, is not enough to place the shuttle into orbit, so for the next two minutes the shuttle begins a shallow descent back toward Earth. During this maneuver, the shuttle increases its speed to 17,500 miles (28,160 kilometers) per hour. At about T-plus 7.7 minutes, the shuttle's main engines throttle down to avoid subjecting the shuttle and its crew to gravitational forces over 3g. (G is the acceleration of gravity at the surface of Earth. It is equal to 32.2 feet [9.8 meters] per second squared.)

By T-plus 8.5 minutes, the shuttle has reached an altitude of about 71 miles (114 kilometers). At this point, all of the fuel in the external tank has been exhausted, and the orbiter's computer shuts down the shuttle's main engines. Thirty seconds later, the external tank separates from the orbiter. The thirty-million-dollar tank breaks up in Earth's atmosphere before falling into either the Pacific Ocean or the Indian Ocean.

The orbiter continues to climb in altitude. At T-plus 10.5 minutes, two small thrusters on the orbiter, a rocket system known as the orbital maneuvering system, fire to place the orbiter in a standard low orbit of 185 miles (300 kilometers) above Earth's surface. If the shuttle mission involved docking with the *Mir* space station or the International Space Station, the thruster would fire again at T-plus 45 minutes to place the orbiter in a higher orbit of 250 miles (400 kilometers).

In orbit, the shuttle circles Earth at 17,500 miles (28,160 kilometers) per hour. Each orbit around the planet takes about ninety minutes, and the crew sees a sunrise or a sunset every forty-five minutes.

Landing

When the mission ends, the astronauts aboard the shuttle perform a number of checklist items, including cleaning up the crew cabin, powering down scientific experiments, and closing the cargo bay doors. The astronauts then don the pressure suits they wore during ascent. After the astronauts are suited and seated, the orbital maneuvering system fires to raise the nose of the orbiter slightly and to reduce its speed. From a point halfway around the world, the shuttle begins its reentry.

As the orbiter enters Earth's atmosphere, drag from the atmosphere begins to slow it down. (As an object moves through the atmosphere, it collides with air particles, which offer resistance and slow down the object. Atmospheric drag decreases with altitude.) The slower the speed of the orbiter, the faster it descends through the atmosphere. About thirty minutes after the firing of the orbital maneuvering system, the orbiter begins to penetrate Earth's atmosphere in earnest. Tremendous heat builds up on the orbiter's underside until it reaches a maximum at twenty minutes before landing. The reentry heat also causes a communications blackout that lasts from twenty-five until twelve minutes before landing.

During the last sixteen minutes before landing, the orbiter performs four S-turn maneuvers like that of a giant slalom skier. Each of these turns removes energy from the vehicle, slowing it down. The last S-turn is performed five minutes prior to landing while the orbiter's speed is still more than 1,500 miles (2,414 kilometers) per hour. At this point, the shuttle is at 83,000 feet (25,300 meters). Its target is a 15,000-foot (4,570-meter) runway at the Kennedy Space Center. (The first nine shuttle spaceflights landed at Edwards Air Force Base in California.)

Eighty-six seconds prior to landing, the orbiter is at an altitude of 13,000 feet (3,960 meters) and traveling at a speed of 425 miles (685 kilometers) per hour. Its rate of descent is

The landing of the orbiter, part of the space shuttle *Columbia*. When the orbiter touches down, it is traveling at a speed of about 215 miles per hour. *(National Aeronautics and Space Administration)*

roughly 22,000 feet (6,700 meters) per minute, compared to an average jet airplane that has a rate of descent of 700 feet (210 meters) per minute. Fourteen seconds prior to touchdown, the orbiter's landing gear is lowered. When the orbiter finally touches down, it is traveling at a speed of 215 miles (345 kilometers) per hour. When all three landing-gear wheels are firmly on the runway, a small drag chute is released to help the wheel brakes slow the orbiter until it finally comes to a stop. The entire landing sequence is done without any power. In a sense, the astronauts are flying nothing more than a large glider.

The program is born

In the mid-1960s, when the Apollo program was well on its way to putting an astronaut on the Moon before the end of the decade, NASA began to develop plans for future space exploration. Officials at the space agency envisioned building a large space station that would be served by reusable space shuttles. These shuttles would also provide services for a permanently manned colony on the Moon and possible manned missions to Mars.

By the 1970s, however, the U.S. public had lost interest in space exploration. Instead, it focused on rising social problems, such as racial discrimination and urban unrest, and the Vietnam War (1954–75). In fact, the increasingly unpopular war had become the single greatest political controversy in the country by 1970. Its enormous financial and human cost were almost staggering. The huge expense of the war fueled inflation (the continuing rise in the general price of goods and services because of an overabundance of available money) and threatened to send the nation into a recession (a period of extended economic decline).

As a result, NASA's budget was cut, severely limiting the projects it could undertake. A large space station was deemed too expensive, but a space shuttle program was deemed feasible if it could provide a low-cost, economical space transportation system. From the beginning, therefore, economics was the dominant aspect of the space shuttle program. Beating the former Soviet Union (present-day Russia) to the Moon had been the goal of the Apollo program, regardless of the cost.

When originally conceived, the shuttle was to have been a two-stage, fully reusable system. The orbiter, a smaller manned winged vehicle, would ride piggyback on the booster, a larger manned winged vehicle. The two stages would be launched like a rocket. Then, at an altitude of about 50 miles (80 kilometers), the two stages would separate. The booster would be flown back to land near the launch site while the orbiter would fire its own engines to place it into orbit. Once the mission was completed, the orbiter would then return to Earth. Under its preliminary design, the orbiter would have been able to carry a 24,915-pound (11,300-kilogram) payload into orbit.

The projected cost for development of the program was initially ten billion dollars. When the U.S. government told NASA officials that the figure was far too high, the agency

Interesting Shuttle Facts

The shuttle's three main engines and two solid rocket boosters produce more than 93.5 times the amount of thrust at liftoff than did the Redstone rocket, the United States's first manned launch vehicle.

Each of the shuttle's three main engines weighs one-seventh as much as a train engine, but one delivers as much horsepower as thirty-nine locomotives.

The energy released by the three main engines is equivalent to the output of twenty-three Hoover Dams.

The space shuttle can accelerate to a speed of more than 17,000 miles (27,353 kilometers) per hour in only about eight minutes.

The two solid rocket boosters produce more thrust at liftoff than the combined thrust of thirty-five 747 jumbo jet airplanes.

The plume of flame coming from the solid rocket boosters at launch ranges up to 500 feet (152 meters) in length.

The speed of the gases exiting the solid rocket boosters is more than 6,000 miles (9,654 kilometers) per hour, which is roughly 8 times the speed of sound at normal atmospheric pressure and 2.5 times the speed of a high-powered rifle bullet.

Filled with propellant, a solid rocket booster is the same height as the Statue of Liberty (without its pedestal), but weighs almost three times as much.

Fuel cells that provide electrical power for systems on the orbiter produce drinking water for the crew as a by-product.

The Remote Manipulator System, the shuttle's robot arm, can move objects in space about the size of a Greyhound bus.

---

redesigned the shuttle. The new design, which featured the orbiter, the external tank, and the solid rocket boosters, came in at a price of five-and-a-half billion dollars. On January 5, 1972, U.S. president Richard M. Nixon (1913–1994) directed NASA to proceed with the development of the space shuttle. Supporters of the program hailed the president's decision, arguing that the shuttle would help restore confidence in the country's technological superiority, both at home and abroad. Critics, however, blasted the program, saying the money should be spent instead on needed social programs.

The shuttle, part spacecraft and part aircraft, required several technological advances. Among these were the development of the thousands of insulating thermal tiles able to stand the heat of reentry over the course of many missions and the sophisticated engines that could be used again and again without being thrown away. NASA hoped that the first shuttle would fly by 1977. It anticipated that a full fleet of four space shuttles would be in complete operation by 1984. Combined, the shuttles were projected to complete twenty-five to sixty missions per year at a cost of ten to twenty million dollars per flight.

None of these goals was met. The first shuttle spaceflight did not take place until 1981, and the fourth space shuttle made its first flight in 1985. The space shuttle program has never delivered on its promise of routine access to space. The most shuttle missions flown in one year, nine, took place in 1985. And the program has been far from economical. Each shuttle flight costs between four hundred million and one billion dollars. At the beginning of the twenty-first century, the operating cost of the shuttle program was more than three billion dollars per year, which is approximately one-quarter of NASA's entire yearly budget.

The debut

The first space shuttle orbiter, known as OV-101, rolled out of an assembly facility in Palmdale, California, on September 17, 1976. The shuttle was originally to be named *Constitution,* but fans of the television show *Star Trek* had started a write-in campaign urging the Nixon administration to name the shuttle after the starship on the show, *Enterprise.* The campaign proved successful.

The *Enterprise* had no engines and was built solely to test the shuttle's gliding and landing ability. Early glide tests, which began in February 1977, were conducted without astronauts and with the orbiter lifted into the air attached to the back of a converted 747 jumbo jet airplane.

Enterprise took to the air on its own on August 12, 1977, when astronauts Fred W. Haise Jr. (1933–) and C. Gordon Fullerton (1936–) flew the 150,000-pound (68,000-kilogram) glider around a course and made a perfect landing. They had separated from the 747 jet at an altitude of 22,800 feet (6,950 meters) and glided to a runway landing at Edwards Air Force Base in California. After its fifth test, *Enterprise* was retired from the program.

The first spaceworthy orbiter, *Columbia*, made its debut flight on April 12, 1981. Aboard were commander John W. Young (1930–) and pilot Robert L. Crippen (1937–). *Columbia*'s only mission was to test its orbital flight and landing capabilities. After spending fifty-four hours in space and completing thirty-six Earth orbits, Young and Crippen brought the shuttle in for a safe landing. During the launch, an overpressure wave created by the solid rocket boosters resulted in the loss of 16 thermal tiles and the damage of 148 more. Upon learning of the problem, NASA engineers immediately made modifications to the boosters to eliminate the wave.

The flight of *Columbia*, known technically as STS-1, marked a new era in human spaceflight. Many believed that within a few years, shuttle flights would take off and land as predictably as airplanes. *Columbia* went on to make four more flights before the second orbiter, *Challenger*, made its first flight on April 4, 1983. The two other original orbiters, *Discovery* and *Atlantis*, made their respective debut flights on August 30, 1984, and October 3, 1985. *Endeavour*, which replaced the ill-fated *Challenger*, first flew on May 7, 1992.

Even though NASA boasted that the shuttle would be able to perform dozens of missions a year with minimal repair, the vehicle was unable to perform under the vigorous standards that were set for it. Space exploration is vastly expensive and NASA is an organization that has long been underfunded by the federal government. Funding for NASA depends on the commitment of the political party that is in power and whatever current domestic and world situations that direct government action. In the 1970s and early 1980s, pressure to balance the budget further eroded NASA's ability to monitor safety and control quality. Before that time, NASA was once the toughest quality-control operation in or out of government. After, it frequently cut corners and sacrificed safety to meet its goals.

In spite of this, the shuttle program accomplished a string of historic successes on the first twenty-four missions between April 1981 and January 1986. The fifth shuttle flight, STS-5, which lifted off on November 11, 1982, was the first operational mission. Astronauts aboard this *Columbia* flight launched two commercial communication satellites. On June

Spacelab, a portable science laboratory built by the European Space Agency, was designed to allow scientist-astronauts to perform a multitude of experiments. *(National Aeronautics and Space Administration)*

18, 1983, Sally K. Ride (1951–) made history when she became the first U.S. female astronaut in space. Her flight came aboard the second mission of the orbiter *Challenger*. In August of that year, on another *Challenger* flight, Guion Bluford Jr. (1942–)

became the first African American astronaut to fly in space by serving on the crew of STS-8. Other memorable flights included those in which members of the U.S. Congress, Senator Jake Garn of Utah and Representative Bill Nelson of Florida, rode aboard shuttle flights in 1985 and 1986, respectively. On other shuttle flights, commercial satellites were not only deployed, but retrieved and repaired in space, as well.

Spacelab, a portable science laboratory, made its first flight into space aboard STS-9 (*Columbia*), which launched on November 28, 1983. Built by the European Space Agency (ESA), the reusable laboratory was designed to allow scientist-astronauts to perform experiments on a wide variety of subjects in microgravity conditions while orbiting Earth. Those subject areas ranged from the life sciences to astronomy to Earth observation to materials science (the study of materials such as metals, glasses, ceramics, polymers, and semiconductors). Spacelab was mounted inside the cargo bay of the orbiter. It consisted of an enclosed pressurized module where the astronauts worked in a shirt-sleeve environment, and smaller U-shaped unpressurized pallets that exposed materials and equipment to space. The pallets carried instruments such as telescopes that could be exposed to space and pointed with high accuracy at stars, the Sun, Earth, or other targets of observation. Spacelab was utilized to some degree on twenty-four shuttle missions between 1983 and 1998.

The Manned Maneuvering Unit (MMU), a one-man propulsion backpack, made its debut during the *Challenger* flight (STS-41B) in February 1984. It allowed astronauts to make the first untethered EVAs, or spacewalks. The MMU snaps onto the back of a spacesuit's portable life-support system. An astronaut wearing the MMU can work outside of the orbiter up to 330 feet (100 meters) away. The unit, which weighs 310 pounds (140 kilograms), is powered by twenty-four thruster jets that burn nitrogen gas. The two pressurized nitrogen tanks provide EVA support for up to six hours at a time.

Mounting pressure and the first disaster

In 1982, in an effort to be more cost-effective, NASA had begun to allow businesses to deliver their payloads into space using the orbiter's cargo bay. However, the U.S. military and certain politicians were opposed to such commercial use of the shuttle. The U.S. government then directed NASA to schedule as many as twenty-four shuttle flights a year, of which the U.S. Air Force would reserve six for its own exclusive military use. The air force, which had previously used rockets to place its satellites into space, decided instead to launch all by shuttle. This put enormous pressure on NASA.

By the end of 1985, NASA had a dismal shuttle flight record, with no more than nine missions having flown in any given year. Greatly underestimating the turnaround time between scheduled launches to be 160 hours, NASA discovered that it needed 1,240 hours minimum. In addition to technical and financial constraints, NASA often found its shuttle schedule to be hampered by weather conditions.

With pressure mounting to meet an impossible schedule, NASA decided that 1986 would be the shuttle program's breakthrough year. In January, it announced an ambitious schedule of fifteen missions using all four of its shuttles. The schedule had to be implemented immediately in order to realize its goal of more than one per month, but technical delays interfered. After at least seven separate postponements, *Columbia* flew the year's first shuttle mission, STS-61C, on January 12. Bad weather prolonged its stay in space, and by the time *Columbia* returned to Earth on January 18, NASA's 1986 schedule was already in jeopardy.

Meanwhile, *Challenger,* which had last flown on November 6, 1985, was being readied for the second January mission. That mission, STS-51L, was to feature the much-publicized Teacher in Space broadcasts as well as plans to launch a Tracking Data and Relay Satellite (TDRS). During the six-day mission, the astronauts would also deploy the Spartan-Halley comet research observatory. Since the observatory had to be launched into orbit no later than January 31, the schedule was tight and inflexible.

The seven-person crew chosen for the mission was commanded by Francis R. "Dick" Scobee (1939–1986), who had piloted a 1984 shuttle mission. This would be the first time in space for his pilot, Michael J. Smith (1945–1986), and for the payload specialist in charge of the TDRS, Gregory B. Jarvis (1944–1986). Mission specialists Judith A. Resnik (1948–1986), Ronald E. McNair (1950–1986), and Ellison S. Onizuka (1946–1986), who ran the satellites and experiments, were all

The seven-person crew of the space shuttle *Challenger*. Back row, left to right: Ellison Onizuka, Christa McAuliffe, Gregory Jarvis, Judith Resnik. Front row, left to right: Michael J. Smith, Francis R. Scobee, Ronald E. McNair. *(AP/Wide World Photos)*

experienced space travelers, having flown on previous shuttle missions. The crew also included another rookie at space travel, teacher Sharon Christa McAuliffe (1948–1986) from Concord High School in New Hampshire.

The Teacher in Space Program was as an extension of NASA's Space Flight Participation Program, which was designed to open space shuttle flights to a broader segment of private citizens. In August 1984, U.S. president Ronald Reagan (1911–1994) had announced that a teacher would be chosen as the first private citizen to fly into space aboard a space shuttle. During the application period, which lasted from December 1984 to February 1985, more than eleven thousand teachers applied. In the summer of 1985, McAuliffe, a high school economics and history teacher, was selected to become the first teacher in space. During the shuttle mission, she was to have conducted two live television teaching lessons. The lessons involved experiments designed to demonstrate Newton's laws and the effects of microgravity on magnetism, among other principles.

Scheduled for January 22, the *Challenger* mission was first postponed to January 24, then to January 25. A forecast of bad weather for January 26 held up the mission until Monday, January 27, when a problem with a hatch bolt suddenly developed. By the time this problem was corrected, crosswinds had built up to a dangerous thirty knots. Although the crew was ready to launch and the shuttle had been fueled, liftoff had to be rescheduled for Tuesday, January 28.

That night, temperatures in Cape Canaveral, Florida, dropped to well below freezing. NASA managers and contractors met for a late-night review. They were becoming increasingly concerned about the cold weather. No shuttle had ever been launched at temperatures below 53°F (12°C). Engineers from Morton Thiokol, a NASA contractor, warned that the O-rings that seal the joints on the shuttle's solid rocket boosters stiffen in the cold and lose their ability to seal properly. NASA managers wanted to know whether the flight could be made. Morton Thiokol managers, overruling their own engineers, signed a waiver stating that the solid rocket boosters were safe for launch at the colder temperatures.

Having decided to go ahead with the launch, NASA turned its attention to the ice on the shuttle and launch pad, which formed as temperatures dipped from 29°F (–1.6°C) to 19°F (–7°C). Icicles actually formed on the shuttle. If they broke off during launch, they could damage the thermal tiles. NASA rescheduled the launch from 9:38 A.M. to 10:38 A.M., and then to 11:38 A.M. Meanwhile, inspection teams surveyed the craft's condition and reported that the ice buildup had caused no apparent abnormalities. Finally, at precisely 11:38 A.M., *Challenger* lifted off.

As *Challenger* rose into a clear, cold blue sky, no one on the ground or in the shuttle realized that a fire flamed out of the right solid booster rocket, jetting down toward the external fuel tank. The shuttle then rolled to align itself on the proper flight path and throttled back its engines. The plume of flame became evident about T-plus 59 seconds. By T-plus 64 seconds, the fire had burned a gaping hole in the casing

Just one second after pilot Michael Smith uttered the words "Uh oh," the space shuttle *Challenger* exploded twenty miles off the coast of Florida. *(AP/Wide World Photos)*

of the booster. At T-plus 72 seconds, it loosened the strut that attached the booster to the external tank. The cockpit flight recorder taped pilot Michael Smith uttering, "Uh oh." This was the only evidence that anyone onboard suspected any trouble.

One second later, the loosened booster slammed into the tip of *Challenger*'s right wing. Then, at an altitude of 46,000 feet (14,020 meters), the booster crashed into the fuel tank and set off a massive explosion. The shuttle was traveling at more than 1,500 miles (2,414 kilometers) per hour.

Challenger exploded 20 miles (32 kilometers) off the coast of Florida. The force of the explosion sent debris flying to an altitude of 20 miles (32 kilometers) above Earth's surface. Burning fragments of the shuttle rained down for the next hour. Of all the accidents in the twenty-five-year history of manned spaceflight, the *Challenger* disaster was by far the worst. It marked the first time that U.S. astronauts had lost their lives during a mission. The disaster, viewed continuously on television, sent shock waves through the nation.

NASA began rescue operations immediately, but the chance of finding survivors was very remote. Once the solid rocket boosters were ignited, the crew had no survivable abort options. If something went wrong at that critical moment of the launch, there was nothing anyone could do.

The Rogers Commission

On February 3, 1986, President Reagan established an independent presidential commission to investigate the accident. He appointed William P. Rogers (1913–2001), who had served as secretary of state under President Nixon, to chair the commission. Joining him on the commission were Neil Armstrong (1930–), the first U.S. astronaut to set foot on the Moon, and a host of scientists and space experts.

Six weeks after the disaster, *Challenger*'s crew cabin was recovered from the floor of the Atlantic Ocean, and the crew members were buried with full honors. Considerable speculation centered on whether the crew had survived the initial explosion. Evidence finally released by NASA indicated that the crew did indeed survive breakup and separation and had initiated emergency procedures. It is unknown if the entire crew remained conscious throughout the two-minute free fall into the ocean, but at least two crew members had activated emergency air packs during that time.

Although the crew cabin has never been exhibited publicly, photographs of the cabin showed nothing recognizable. Experts estimate that the module hit the surface of the ocean at a speed of nearly 2,000 miles (3,218 kilometers) per hour. The sixteen-foot-high (five-meter-high) cabin was compressed into a solid mass half its original size, which would certainly have killed anyone still alive in the module. The cabin's thick windows were shattered, but there was no evidence of fire. In fact, the fireball of the explosion did not cause the destruction of *Challenger;* instead, severe aerodynamic loads created by the external fuel tank explosion broke the shuttle apart.

Space Shuttle Landmarks

Mission	Shuttle	Launch	Highlights
STS-1	*Columbia*	April 12, 1981	First STS flight
STS-5	*Columbia*	November 11, 1982	First STS operational mission
STS-7	*Challenger*	June 18, 1983	First spaceflight by a U.S. female astronaut (Sally K. Ride)
STS-8	*Challenger*	August 30, 1983	First spaceflight by an African American male astronaut (Guion Bluford Jr.); first STS night launch and landing
STS-41B	*Challenger*	February 3, 1984	First untethered spacewalks
STS-41C	*Challenger*	April 6, 1984	First in-flight capture, repair, and redeployment of an orbiting satellite
STS-41G	*Challenger*	October 5, 1984	First spacewalk by a U.S. female astronaut (Kathryn Sullivan)
STS-61C	*Columbia*	January 12, 1986	First spaceflight by an Hispanic male astronaut (Franklin R. Chang-Dìaz)
STS-51L	*Challenger*	January 28, 1986	First STS in-flight accident; first teacher in space (Christa McAuliffe)
STS-47	*Endeavour*	September 20, 1992	First spaceflight by an African American female astronaut (Mae Jemison)

The Rogers Commission released its findings on June 6, 1986. It determined that the immediate physical cause of the *Challenger* disaster was a failure in the joint between the two lower segments of the right solid rocket booster. Rubber O-rings seal the joints between the booster's four sections. Zinc chromate putty keeps the hot combustion gases inside the booster from coming into contact with the rubber rings.

When it checked into the history and performance of this O-ring sealing system, the Rogers Commission was shocked to learn that the O-rings had failed regularly, even if only partially, on previous shuttle flights. Although concerned about the frailty of the seals, NASA and Morton Thiokol decided not

Mission	Shuttle	Launch	Highlights
STS-56	*Discovery*	April 8, 1993	First spaceflight by an Hispanic female astronaut
STS-61	*Endeavor*	December 2, 1993	First STS mission to service the Hubble Space Telescope
STS-63	*Discovery*	February 3, 1995	First female STS pilot (Eileen Collins)
STS-71	*Atlantis*	June 27, 1995	First STS docking with Mir space station
STS-95	*Discovery*	October 29, 1998	Oldest astronaut to fly in space (John Glenn)
STS-88	*Endeavour*	December 4, 1998	First STS assembly flight of International Space Station
STS-93	*Columbia*	July 23, 1999	First female STS mission commander (Eileen Collins)
STS-92	*Discovery*	October 11, 2000	One-hundredth STS mission, including one-hundredth spacewalk in U.S. space program
STS-113	*Endeavor*	November 23, 2002	First spaceflight by a Native American astronaut (John Herrington)
STS-107	*Columbia*	January 16, 2003	Second STS in-flight accident

to redesign the system. Because the seals had never failed completely, both had considered the O-rings to be an acceptable risk. But the cold temperatures on the day of *Challenger*'s flight made the O-rings less flexible than normal, and they did not completely seal the joint. Photographs reveal that even before the shuttle had cleared the launch tower, hot gas was already escaping by the O-rings.

The 256-page report of the Rogers Commission concluded that NASA made a grave mistake in its decision to launch *Challenger*. The commission blamed the management structure of both NASA and Morton Thiokol for not allowing critical information to reach the proper people. A U.S. congressional committee, which spent two months conducting its own hearings, agreed with this assessment. The committee determined that the technical problem had actually been recognized early enough to prevent the disaster, but that NASA had placed a higher priority on meeting flight schedules and cutting costs than on flight safety.

After the reports had been released, the nation's confidence in NASA was badly shaken. Astronauts were especially disturbed. They had never been consulted or even informed about the dangers to which they were exposed by the current sealing system. Allowing astronauts a greater role in approving launches was one of the nine recommendations the Rogers Commission made to NASA. The commission's other recommendations included a complete redesign of the solid rocket booster joints, the development of an escape system that would allow astronauts to leave the shuttle while in flight in some cases, and a sweeping reform of the shuttle program's management structure to allow improved communication between engineers and managers.

The *Challenger* disaster grounded the shuttle fleet for more than two-and-one-half years while the required improvements were made to the remaining orbiters. During this time, several key people, including a number of experienced astronauts, left NASA. They were disillusioned with the space agency and frustrated that there might be even fewer chances to fly.

A return to flight

On September 29, 1988, with the launch of *Discovery* (STS-26), NASA inaugurated a new era of space shuttle operations. Learning from one of its greatest tragedies, it adopted a more relaxed pace, averaging about eight launches per year. NASA was able to rebuild and maintain a space shuttle program that was remarkably safe and reliable, for a while.

More than two hundred safety improvements and modifications had been made to the shuttle fleet. The improvements included a major redesign of the solid rocket boosters, the addition of a crew escape and bailout system, stronger landing gear, more powerful flight control computers, and updated navigational equipment.

Shuttle improvements did not stop with *Discovery*. *Endeavour*'s first flight (STS-49) on May 7, 1992, unveiled many

The design of space shuttle *Endeavour* showed many improvements, including a drag chute to assist braking during landing, improved steering, and more reliable power hydraulic units. *(National Aeronautics and Space Administration)*

improvements, including a drag chute to assist braking during landing, improved steering, and more reliable power hydraulic units. Further upgrades to the shuttle system occurred when *Columbia* was modified for its June 25, 1992, flight (STS-50) to allow long-duration flights. The modifications to the orbiter included an improved toilet, a system to remove carbon dioxide from the air, connections for a pallet of additional hydrogen and oxygen tanks to be mounted in the cargo bay, and extra stowage room in the crew cabin.

With these improvements, the shuttles returned to their former status as the workhorses of space exploration. Early in the shuttle program, communications satellites were common payloads, with as many as three delivered into orbit on the same mission. The *Challenger* accident led to a change in that policy. After returning to flight in the fall of 1988, shuttles carried only those payloads unique to the shuttle program or those that require a human presence. The majority of those were scientific and defense missions.

Highlights of scientific missions undertaken since the *Challenger* accident include those that have launched spacecraft to study other celestial objects in the solar system. On May 4, 1989, the shuttle *Atlantis* (STS-30) carried aloft the *Magellan* spacecraft, the first planetary explorer to be launched by a space shuttle. Magellan's mission was to make the most highly detailed map of Venus ever captured. It completed that mission during the four years it orbited the planet between 1990 and 1994. Four months later, *Atlantis* (STS-34) helped launch another spacecraft: *Galileo*. It carried out the first studies of Jupiter's atmosphere, moons, and magnetosphere (the region of space around a celestial object that is dominated by the object's magnetic field) from orbit around the planet. The *Ulysses* probe, which studied the Sun, the makeup of the solar wind, and the interplanetary medium (the space between planets including forms of energy and dust and gas), went into space aboard the shuttle *Discovery* (STS-41), which launched on October 6, 1990.

Three of NASA's four Great Observatories—the Hubble Space Telescope, the Compton Gamma Ray Observatory, and the Chandra X-ray Observatory—were placed into orbit during shuttle missions. Hubble was the first, having been deployed from the shuttle *Discovery* (STS-31) on April 25, 1990. Compton, placed in orbit on April 7, 1991, by the astronauts aboard *Atlantis* (STS-37), was the heaviest astrophysical payload ever flown up to that time. It weighed 34,442 pounds (15,620 kilograms). Chandra was carried aloft by the shuttle *Columbia* (STS-93) and deployed on July 23, 1999.

Spacehab, a small commercially built laboratory module similar in concept to Spacelab, was first carried into space aboard the June 1993 flight of the shuttle *Endeavour* (STS-57). Like Spacelab, Spacehab provided extra working space in which astronauts were able to carry out experiments.

A vision for the future

In April 1996 NASA began a four-phase plan, called the Space Shuttle Upgrade Program, to keep the existing space shuttle fleet healthy and flying through at least the year 2012. The plan also proposed modifications and upgrades that NASA hoped might keep the fleet flying through the year 2030.

Phase one of the plan called for improvements to the space shuttle that were necessary to allow it to support construction and maintenance of the International Space Station, which was the chief program goal at the beginning of the twenty-first century.

Phase two called for improvements in ground operations to decrease the amount of time it took to service and maintain shuttles between flights. The goal was to provide support for an average of fifteen launches per year.

Phase three called for a number of modifications to the onboard systems of the orbiters that NASA hoped would also result in decreased processing and maintenance time. More ambitious elements of this plan called for completely replacing toxic fuels with nontoxic fuels in key orbiter systems.

Finally, phase four called for the significant redesign of the space shuttle fleet and its basic configuration. An interesting proposal in this phase was the introduction of a booster that would fly back to the launch site and save precious servicing time.

A future interrupted by another disaster

By the beginning of the twenty-first century, space travel seemed commonplace yet again. The original purpose of the space shuttle program, to ferry supplies to a space station, was finally being realized as shuttle missions were visiting the International Space Station (ISS) on average every few months. Then on September 11, 2001, U.S. public attention was fixated on New York City and Washington, D.C., as commercial passenger jets hijacked by terrorists crashed into the World Trade Center and the Pentagon. Space exploration suddenly no longer seemed important.

Yet NASA continued with its vision for the future, flying shuttle missions to the ISS. One mission whose flight was not aimed at the space station was STS-107, which launched from the Kennedy Space Center on January 16, 2003. The astronauts aboard *Columbia* during this sixteen-day flight performed what many thought were routine scientific chores: These included an examination of dust in the Middle East; a study of the planet's ozone layer; experiments designed by schoolchildren in six countries to observe the effects of microgravity on spiders, silkworms, and other creatures; and the extraction of essential oils from rose and rice flowers. In all, the crew completed approximately eighty experiments during the flight.

The flight was commanded by Rick Husband (1957–2003), who had flown on one previous shuttle mission. The mission's pilot was William McCool (1961–2003). The three mission specialists on the flight were Kalpana Chawla (c. 1961–2003), David Brown (1956–2003), and Laurel Clark (c. 1961–2003). Michael Anderson (1959–2003) was the mission's payload commander, while Ilan Ramon (1954–2003) was its payload specialist. Ramon, a colonel in the Israel Air Force, was the first Israeli astronaut to fly into space.

The mission was so routine that U.S. Air Force radars did not track *Columbia*'s reentry into Earth's atmosphere in the early morning of February 1. The shuttle was supposed to touch down at the Kennedy Space Center at 9:16 A.M. EST.

At 8:53 A.M., as the shuttle crossed the California coast at about 15,000 miles (24,135 kilometers) per hour at an altitude of 230,000 feet (70,100 meters), sensors on the shuttle began showing signs of trouble. Data from four temperature indicators on the hydraulic systems on the left side of the vehicle were lost. Because the shuttle seemed to be functioning normally otherwise, ground controllers did not alert the crew. Five minutes later, data was lost from three temperature sensors imbedded in *Columbia*'s left wing. At 8:59 A.M., data was lost from tire temperature and pressure sensors on the shuttle's left side. One of the sensors alerted the crew, who acknowledged the alert when communication with ground control was lost. About one minute later, all data from the shuttle was lost. At the time, *Columbia* was at about 205,000 feet (62,484 meters) over north-central Texas and was traveling at about 13,100 miles (21,080 kilometers) per hour. For several minutes, NASA officials tried to reestablish communication, but they were unsuccessful.

The reason they could not was that *Columbia,* the oldest in NASA's shuttle fleet, had disintegrated in the atmosphere. Debris from the shuttle, including the remains of the seven astronauts onboard, fell across an area of 28,000 square miles (72,520 square kilometers) from north of Dallas, Texas, to western Louisiana. More than twelve hundred shuttle pieces alone were found in Nacogdoches, Texas. The shuttle parts ranged in size from tiny shards to 8-foot (2.4-meter) chunks. Residents in north-central Texas later reported hearing a large boom and seeing smoke trails in the clear sky at the time of the accident.

The *Columbia* disaster had come just a few days after the seventeenth anniversary of the *Challenger* explosion and the thirty-sixth anniversary of a launch pad fire that had claimed the lives of three Apollo astronauts.

European Space Shuttle

In May 2004 the European Space Agency (ESA) successfully tested an unmanned prototype (model) for its European space shuttle. The EADS Phoenix, a German-designed vehicle, was dropped from a helicopter at an altitude of 7,900 feet (2,408 meters). After a ninety-second flight, it made a perfect landing on the test runway located 770 miles (1,240 kilometers) north of the Swedish capital of Stockholm.

The Phoenix prototype shuttle is just less than 23 feet (7 meters) long and weighs 2,640 pounds (1,200 kilograms). It has a wingspan of 13 feet (4 meters). The actual planned shuttle is to be six times the size of the prototype. The ESA hopes that the shuttle, which will be used to send astronauts into space, will be finished sometime between 2015 and 2020.

CAIB

NASA officials immediately grounded the shuttle fleet. Following guidelines established after the loss of *Challenger,* an independent investigating board was created right after the accident. The Columbia Accident Investigation Board, or CAIB, consisted of expert military and civilian analysts who investigated the accident in great detail. Chairing the board was retired U.S. Navy admiral Harold W. Gehman Jr. (1942–). Among those on the thirteen-member board was former astronaut Sally Ride.

It was already known that on liftoff, a piece of foam insulation covering the external fuel tank had broken loose and had hit the underside of *Columbia*'s left wing. A few weeks after the accident, NASA released a volley of e-mails that space shuttle engineers had sent the day before *Columbia* broke up. In those e-mails, the engineers had expressed their concern that the shuttle's left wing might burn off and lead to the complete loss of the orbiter. Their concerns, however, were never forwarded to top management personnel at NASA, who had determined earlier that there was no landing risk.

On August 26, 2003, the CAIB released its 248-page final report on the accident. The board concluded that, indeed, a two-pound chunk of insulating foam from the shuttle's external fuel tank ripped away during its launch and hit a seal on the leading edge of the left wing. The strike created a slit large enough in the reinforced carbon material to let in superheated air that progressively melted the aluminum structure of the left wing as *Columbia* reentered Earth's atmosphere on its return home.

The CAIB was especially critical of NASA's management structure, which the board felt had as much to do with the accident as did the foam. It believed that NASA managers allowed unsafe practices to develop, then quieted discussions regarding solutions to possible problems. Among other recommendations to the space shuttle program, the CAIB directed NASA to:

- Continue the space shuttle program with adequate funding.
- Build a replacement for the shuttle.
- Prevent the shuttle's external fuel tank from shedding any debris before flying again.
- Improve the shuttle's ability to sustain minor debris damage and develop tests to determine the resistance of current materials used in the orbiter.
- Develop the capability to inspect and make emergency repairs to the thermal tiles while the shuttle is in orbit.
- Upgrade the imaging system to provide more useful views of the shuttle during liftoff. Also consider using aircraft to provide additional views of the orbiter during ascent.

- Design a better system to collect sensor data from the craft.

- Expand a training program for NASA mission teams to look beyond launch and ascent, including the potential for loss of the shuttle and crew while in orbit.

- Establish an independent technical engineering authority that looks at safety and does not have responsibility for schedule or program costs.

- Reformulate management so that NASA's main office of safety has independent oversight over shuttle safety.

- Conduct a vehicle recertification of the shuttle and its systems before operating the craft beyond 2010.

A changed future in space

In the months following the *Columbia* accident, polls revealed that support for the space program remained strong among the U.S. public. Two-thirds of those polled believed that the space shuttle should continue to fly, and nearly three-quarters said that the space program was a good investment.

But on January 14, 2004, U.S. president George W. Bush (1946–) outlined a new course for U.S. space exploration. He proposed to keep spending several billion dollars a year to put the remaining three shuttles back into space to finish ferrying parts to the International Space Station. Once construction of the station was completed, the entire shuttle fleet would be retired by 2010. The president then unveiled a plan to develop a new manned exploration vehicle, one that would lead manned missions back to the Moon sometime between 2015 and 2020. Once a permanent lunar base was established, it could be used as a stepping-stone for future manned trips to Mars.

President Bush's vision for future space exploration received a lukewarm response. Some thought the president's proposal was a wake-up call for NASA, helping give it a clear direction for the twenty-first century. Others, however, believed it was beyond the realm of what the space agency could accomplish. They also felt that the price of such an undertaking was too high and wasteful.

By mid-2004, NASA believed that it had made sufficient progress to return the shuttle fleet to operation by the spring of 2005. It had originally hoped to place shuttles back in orbit sooner, but the agency felt that it needed the extra time to resolve persistent technical problems with the shuttles, such as preventing the external fuel tank from shedding foam insulation, and to prepare a potential backup shuttle that could be used in case a rescue is needed for a shuttle already in orbit.

For More Information

Books

Cole, Michael D. *The Columbia Space Shuttle Disaster: From First Liftoff to Tragic Final Flight.* Revised ed. Berkeley Heights, NJ: Enslow, 2003.

Holden, Henry M. *The Tragedy of the Space Shuttle Challenger.* Berkeley Heights, NJ: MyReportLinks.com Books, 2004.

Jenkins, Dennis R. *Space Shuttle: The History of the National Space Transportation System.* Third ed. Cape Canaveral, FL: D. R. Jenkins, 2001.

Reichhardt, Tony, ed. *Space Shuttle: The First 20 Years—The Astronauts' Experiences in Their Own Words.* New York: DK Publishing, 2002.

Ride, Sally. *To Space and Back.* New York: HarperCollins, 1986.

Web Sites

"The Challenger Accident." *Federation of American Scientists Space Policy Project.* http://www.fas.org/spp/51L.html (accessed on August 19, 2004).

"Human Space Flight: Space Shuttle." *National Aeronautics and Space Administration.* http://spaceflight.nasa.gov/shuttle/ (accessed on August 19, 2004).

"Remembering *Columbia STS-107.*" *National Aeronautics and Space Administration.* http://history.nasa.gov/columbia/index.html (accessed on August 19, 2004.

"Space Shuttle." *NASA/Kennedy Space Center.* http://www.ksc.nasa.gov/shuttle/ (accessed on August 19, 2004).

"Space Shuttle Mission Chronology." *NASA/Kennedy Space Center.* http://www-pao.ksc.nasa.gov/kscpao/chron/chrontoc.htm (accessed on August 19, 2004).

12

Ground-based Observatories

The exploration of space is not limited to the flights of astronauts aboard spacecraft and shuttles launched into space by rockets and boosters. That type of space exploration has a history that extends back only to the mid-twentieth century. At the farthest, humans have traveled only about 252,780 miles (406,720 kilometers) away from Earth—that's the distance to the Moon at its apogee (pronounced AP-eh-gee), or farthest point of its orbit. (The distance between the Moon and Earth varies because the Moon's orbit around Earth is elliptical, or oval-shaped. On average, it is located at a distance of 238,900 miles [384,390 kilometers].) Astronauts aboard space shuttles and space stations have stayed relatively close to Earth, conducting work in space at a distance of about 185 to 250 miles (300 to 400 kilometers) above the planet's surface.

The deep exploration of space has come through astronomical observations, or the study of the sky. For centuries, astronomers have used telescopes to observe the Moon, the other planets in the solar system, asteroids and comets, stars at the far reaches of the Milky Way (the galaxy that contains our solar system), and extremely bright and distant objects

The 200-inch-diameter Hale Telescope is housed at the Palomar Observatory in California. (© Bettmann/Corbis)

known as quasars (pronounced KWAY-zarz). From there, astronomers have gone on to investigate the formation of the planets and Sun, the life cycles of stars, and the age and formation of the universe.

Astronomy is not just about visible light. What the human eye sees is only a small portion of the activities and processes underway in the universe. Astronomers study the universe by measuring electromagnetic radiation emitted by planets, stars, galaxies, and other distant celestial objects. Radiation is the emission and movement of waves or atomic particles through space or other media. Electromagnetic radiation is radiation that transmits energy through the interaction of electricity and magnetism. When astronomers view the night sky through forms of electromagnetic radiation, they see an entirely different picture: Hot gases seethe and boil when viewed at infrared wavelengths, newly forming galaxies and stars glow with X rays, and mysterious objects generate explosive bursts of gamma rays.

The various forms of electromagnetic radiation, including gamma rays, X rays, optical and infrared radiation, and radio waves, move through space in waves. Like any wave, they can be described by two properties: wavelength and frequency. The wavelength is the distance between one crest, or peak, of a wave and the next corresponding peak. Frequency is the rate at which two successive identical parts of the wave pass a given point. Wavelength and frequency have a reciprocal relationship with each other: As one increases, the other must decrease.

Gamma rays are short-wavelength, high-energy radiation formed either by the decay of radioactive elements or by nuclear reactions. X rays, which have wavelengths just shorter than ultraviolet radiation but longer than gamma rays, can penetrate solids and produce an electrical charge in gases. (Ultraviolet radiation has a wavelength just shorter than the violet, or shortest wavelength, end of the visible light spectrum.) Optical radiation is visible light, or electromagnetic radiation that is detectable by the human eye. The different colors of light the human eye can see correspond to different wavelengths: Red light has the longest wavelength, violet the shortest. Orange, yellow, green, blue, and indigo are in between (moving from red to violet). Infrared radiation is

Words to Know

Apogee: The point in the orbit of an artificial satellite or the Moon that is farthest from Earth.

Astronomy: The scientific study of the physical universe beyond Earth's atmosphere.

Big bang theory: The theory that explains the beginning of the universe as a tremendous explosion from a single point that occurred about thirteen billion years ago.

Cepheid variable: A pulsating star that can be used to measure distance in space.

Chromatic aberration: Blurred coloring of the edge of an image when visible light passes through a lens, caused by the bending of the different wavelengths of the light at different angles.

Concave lens: A lens with a hollow bowl shape; it is thin in the middle and thick along the edges.

Constellation: One of eighty-eight recognized groups of stars that seems to make up a pattern or picture on the celestial sphere.

Convex lens: A lens with a bulging surface like the outer surface of a ball; it is thicker in the middle and thinner along the edges.

Dark matter: Virtually undetectable matter that does not emit or reflect light and that is thought to account for 90 percent of the mass of the universe, acting as a "cosmic glue" that holds together galaxies and clusters of galaxies.

Electromagnetic radiation: Radiation that transmits energy through the interaction of electricity and magnetism.

Focus: The position at which rays of light from a lens converge to form a sharp image.

Galaxy: A huge region of space that contains billions of stars, gas, dust, nebulae, and empty space all bound together by gravity.

Gamma rays: Short-wavelength, high-energy radiation formed either by the decay of radioactive elements or by nuclear reactions.

Infrared radiation: Electromagnetic radiation with wavelengths slightly longer than those of visible light.

Interferometer: A device that uses two or more telescopes to observe the same object at the same time in the same wavelength to increase angular resolution.

Light-year: The distance light travels in the near vacuum of space in one year, about 5.88 trillion miles (9.46 trillion kilometers).

Neutron star: The extremely dense, compact, neutron-filled remains of a star following a supernova.

Observatory: A structure designed and equipped to observe astronomical phenomena.

Pulsar: A rapidly spinning, blinking neutron star.

Quasars: Extremely bright, star-like sources of radio waves that are found in remote areas of space and that are the oldest known objects in the universe.

Radiation: The emission and movement of waves or atomic particles through space or other media.

Radio waves: The longest form of electromagnetic radiation, measuring up to 6 miles (9.7 kilometers) from peak to peak in the wave.

Redshift: The shift of an object's light spectrum toward the red end of the visible light range, which is an indication that the object is moving away from the observer.

Reflector telescope: A telescope that directs light from an opening at one end to a concave mirror at the far end, which reflects the light back to a smaller mirror that directs it to an eyepiece on the side of the telescope.

Refractor telescope: A telescope that directs light waves through a convex lens (the objective lens), which bends the waves and brings them to a focus at a concave lens (the eyepiece) that acts as a magnifying glass.

Solstice: Either of the two times during the year when the Sun, as seen from Earth, is farthest north or south of the equator; the solstices mark the beginning of the summer and winter seasons.

Sunspot: A cool area of magnetic disturbance that forms a dark blemish on the surface of the Sun.

Supernova: The massive explosion of a relatively large star at the end of its lifetime.

Telescope: An instrument that gathers light or some other form of electromagnetic radiation emitted by distant sources, such as celestial bodies, and brings it to a focus.

Ultraviolet radiation: Electromagnetic radiation of a wavelength just shorter than the violet (shortest wavelength) end of the visible light spectrum.

X rays: Electromagnetic radiation of a wavelength just shorter than ultraviolet radiation but longer than gamma rays that can penetrate solids and produce an electrical charge in gases.

electromagnetic radiation with wavelengths slightly longer than those of visible light. Humans cannot see infrared radiation, but can sense its energy as heat on the skin. Radio waves are the longest form of electromagnetic radiation. Their wavelengths measure up to 6 miles (9.7 kilometers) from peak to peak.

Earth's atmosphere provides an effective filter for many forms of cosmic radiation. This condition is crucial for the survival of humans and other life-forms on the planet. The atmosphere blocks gamma rays and X rays, and these forms of radiation must be studied by telescopes launched into space. Optical and infrared radiation and radio waves are able to pass through Earth's atmosphere, although carbon dioxide and water in the atmosphere absorb much of the infrared radiation. Ground-based observatories, structures designed and equipped to observe astronomical phenomena, are thus able to study these forms of electromagnetic radiation.

Astronomers make use of ground-based observatories whenever possible. Although space-based observatories are not affected by the distorting effects of the atmosphere, allowing them to capture incredibly detailed images of the cosmos, they are extremely costly. It is about one thousand times cheaper to build a telescope of a given size on the ground than to launch it into space.

Ancient observatories

From at least the beginning of civilization, ancient humans looked at the stars in the night sky and struggled to make sense of what they saw. Initially, they tried to connect the stars to the world around them by visualizing animals and mythological characters in the constellations they perceived (a constellation is a recognized group of stars that seems to make up a pattern or picture in the night sky). At some point, these ancient humans turned from noting a single pattern or celestial event to making the kinds of repeated observations that could be applied to predict events in their own lives, such as knowing when to harvest crops. Although any written records of ancient celestial observations have been lost to history, some of the physical signs of those activities remain. Among the most intriguing are the sites that, to a modern eye, could have been used as very early observatories.

Between about 3,500 and 1,500 B.C.E., ancient humans in present-day Great Britain and northwestern France marked in the existing landscape thousands of sites that may have been used for astronomical observations. A common arrangement consisted of a natural indicator on the horizon, such as a notch in a mountain. This is known as the foresight. It was aligned with a manmade marking, such as a standing post or stone, or a hollowed-out depression in a rock. This second indicator is known as the backsight. Because of the great distances between some of these foresights and backsights, up to 28 miles (45 kilometers), present-day scientists do not believe that they could have been used to establish the date of a major celestial event, such as a solstice, with certainty. However, it is possible that they were used in Sun-worshipping ceremonies.

Following these simple observing stations that made use of existing sight lines and horizon marks came more recognizable and complex structures. Some astronomers believe that Stonehenge, built over a period of time beginning more than 4,000 years ago on a flat plain in present-day southern England, could have been used to observe the summer solstice, the time when the rising of the Sun is farthest north, and the extreme rising and setting positions of the Moon. Its primary purpose seems to have been ceremonial.

Other cultures in various parts of the world appear to have followed observing practices like those conducted at Stonehenge. Ancient Native Americans in the American Southwest relied on wall calendars, which were created as sunlight penetrated an opening in a house or residential cave to fall on the opposite wall. They used wall calendars to track the motions of the Sun, the Moon, stars, and other celestial events.

Elaborate astronomical observatories were built by the Maya, native people of present-day Central America and southern Mexico, and the Inca, native people of present-day Peru. Mayan culture reached its peak roughly 1,000 years ago. In cities such as Chichén Itzá, they built spectacular temples in which they conducted elaborate ceremonies and made observations of solar equinoxes and solstices. The Inca, who established an empire that lasted from 1100 to about 1500, created similar temples in the ancient cities of Machu Picchu (pronounced MAH-choo PEE-choo) and Llactapata (pronounced yak-tah-PAH-tah), among others.

The telescope

The celestial events viewed at these ancient observatories all had one thing in common: They were seen only with the naked eye. Although sight is perhaps the strongest and most important of humans' special senses (80 percent of all information received by the human brain comes from the eyes), it is limited. The ability of the human eye to work with the human brain to transform light waves into visual images depends upon the shape of the eye and the distance of the object viewed. If any part of the eye is abnormal in shape, vision is reduced. The farther away an object, the harder it is to see. Atmospheric conditions and the amount of available light also affect sight.

What increased humans' vision of the world around them and the sky above was the telescope. The telescope is an instrument that gathers light or some other form of electromagnetic radiation emitted by distant sources, such as celestial bodies, and brings it to a focus. The most common present-day type is the optical telescope, which uses a collection of lenses or mirrors to magnify the visible light emitted by a distant object. There are two basic types of optical telescopes: the refractor and the reflector. The one characteristic that all telescopes have in common is the ability to make distant objects appear to be closer.

The first extension of one of humans' special senses, the telescope shifted authority in the observation of nature from humans to instruments. In the process, it became the predecessor of modern scientific instruments. Yet, it was not the invention of scientists. Rather, it was the product of craftsmen.

The ability of convex (curving outward) and concave (curving inward) transparent material to magnify or minimize images was known in ancient times. However, lenses as they are known today were introduced in Western Europe at the end of the thirteenth century. In the major Italian glassmaking centers of Venice and Florence, techniques for grinding and polishing glass had reached a high state of development. Magnifying glasses became common, but they were cumbersome, especially for reading and writing.

Craftsmen in Venice then began making small disks of glass, convex on both sides, that could be worn in a frame—the first eyeglasses. The shape of these small disks of glass reminded people of the small, flat beans known as lentils.

Because of this, they became known as lenses, which is derived from the Latin word for lentils.

Lenses that were concave on both sides were then introduced in the middle of the fifteenth century. With this development, all of the main ingredients of a telescope—convex and concave mirrors and lenses—were present. Nonetheless, the first optical telescope would not be created for another 150 years.

There is much confusion and debate concerning the origin of the first telescope. Many notable individuals appear to have simultaneously and independently discovered how to make a telescope during the last months of 1608 and the early part of 1609. Regardless of its origins, the invention of the telescope has led to great progress in the field of astronomy.

Contrary to popular belief, the Italian mathematician and astronomer Galileo Galilei (pronounced ga-lih-LAY-oh ga-lih-LAY-ee; 1564–1642) may not have been the first person to use this instrument in astronomy. Instead, that honor may be bestowed on a contemporary of Galileo, English mathematician Thomas Harriot (1560–1621). Harriot developed a map of the Moon several months before Galileo began making observations. Nevertheless, Galileo distinguished himself in the field through his patience, dedication, insight, and skill.

Hans Lippershey, a Dutch lens-grinder, created the first optical telescope in 1608. *(© Bettmann/Corbis)*

The actual inventor of the telescope may never be known with certainty. Its invention may have been an accidental occurrence when some spectacle, or eyeglass, maker happened to look through two lenses at the same time. Many accounts report that Dutch lens-grinder Hans Lippershey (sometimes spelled Lipperhey; 1570–1619) had two lenses set up in his spectacle shop to allow visitors to look through them to see the steeple of a distant church. There is even a story that Lippershey's children actually created the first telescope while they were playing with flawed lenses in his shop.

It is known that the first telescopes were shown in the Netherlands. Records indicate that Lippershey, who thought that his device might be useful in warfare, applied for a patent with the national government of the Netherlands in October 1608. Other Dutch spectacle makers applied for similar patents at the same time. All of these devices consisted of a convex and a concave lens mounted in a tube. The combination of the two lenses magnified objects by three or four times. However, the government of the Netherlands considered the devices too easy to copy to justify awarding any patents.

News of the invention of the telescope spread rapidly throughout Europe. Within a few months, simple telescopes, called "spyglasses," could be purchased at spectacle-makers' shops in Paris. By early 1609 four or five telescopes had made it to Italy. By August 1609 Harriot had observed and mapped the Moon with a six-power telescope.

Galileo's discoveries

Despite Harriot's honor as the first telescopic astronomer, it was Galileo who made the telescope famous. Although there is no evidence that he actually saw one of the telescopes known to be in Italy at the time, he somehow learned of the newly invented instrument. Over several months in 1609 and 1610, Galileo made several progressively more powerful and optically superior telescopes using lenses he ground himself. He then used these telescopes for a systematic study of the night sky. Among his many observations, he saw mountains and craters on the Moon, discovered four moons of Jupiter, viewed sunspots (cool areas of magnetic disturbance that form dark blemishes on the surface of the Sun), and found that the Milky Way consisted of clouds of individual stars.

Galileo summarized his discoveries in a small book titled *Sidereus Nuncius* (*Starry Messenger*), which was published in March 1610. Others working at around the same time claimed to have made similar discoveries—others certainly observed sunspots—but Galileo was first to write about these observa-

Galileo Galilei made several progressively more powerful and optically superior telescopes using lenses he ground himself. He then used these telescopes for a systematic study of the night sky. *(The Library of Congress)*

tions. Consequently, he is generally credited with their discovery.

The observation of the four moons in orbit around Jupiter was especially important to Galileo. This discovery conclusively contradicted the prevailing belief that Earth was at the center of the solar system, with everything in the system revolving around it.

Galileo apparently had no real knowledge of how the telescope worked, but he immediately recognized its value. He set about building an early version of what is now referred to as a refractor telescope. With this type of device, light waves from a distant object enter the top of the telescope through what is called an objective lens. This lens is convex—thicker in the middle than at the edges. As light waves pass through it, they are refracted (bent) so that they converge (come together) at a single point, known as the focus. The distance between the objective lens and the focus is called the focal length. A concave lens, known as the eyepiece, placed at the focus then magnifies the image for viewing.

The first person to give a concise theory of how light passes through the telescope and forms an image was German astronomer Johannes Kepler (1571–1630). He also discussed the various ways in which the lenses could be combined in different optical systems, improving on Galileo's design. Kepler's design used convex lenses for both the objective lens and the eyepiece.

Unfortunately, telescopes built following Kepler's design were not practical for military or everyday use because they inverted (turned upside down) and reversed the image seen through the eyepiece. However, because they offered a greater degree of magnification, a brighter image, and a wider angle of view, they were best for astronomical observations where the inverted image made no difference.

As refractor telescopes came into wider use, though, it became apparent that they had a great defect. The main problem with these early telescopes was the low quality of their glass and the poor manner in which their lenses were ground. However, even the best lenses had an inherent imperfection. Like a prism, a lens bends the different wavelengths (colors) that make up visible light through different angles. The objective lenses in these telescopes did not bend all wavelengths equally. This resulted in the red part of the visible light being brought to a focus at a greater distance from the objective lens. An image of a star viewed through a refractor telescope from this period seemed to be surrounded by colored fringes. This defect is called chromatic aberration. Early astronomers tried to correct this problem by increasing the focal length, but the new instruments were very clumsy to use.

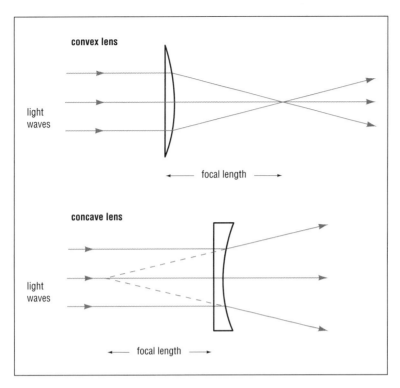

convex lens

light waves

focal length

concave lens

light waves

focal length

Convex and concave lenses were used by astronomers such as Galileo and Johannes Kepler to formulate theories on how light passes through a telescope and forms an image. *(The Gale Group)*

A solution to this problem came in 1729 when English scientist Chester Moor Hall (1703–1771) devised the achromatic lens: a grouping of two lenses, made of different kinds of glass and shape, set close together. As light passes through, the second lens cancels out the false color brought about by the first lens. Hall went on to create the achromatic telescope in 1733. Twenty-five years later, English optician John Dollard improved on the achromatic lens by combining two or more lenses with varying chemical compositions to minimize the effects of aberration.

Newton and the reflector telescope

During the 1680s, English physicist and mathematician Isaac Newton (1642–1727) had begun trying to unravel the problem of chromatic aberration. While observing a beam of sunlight passing through a glass prism, he saw that the beam was split into a rainbow of colors. On the basis of this and other experiments, he decided (incorrectly, it turns out) that the refractor telescope could never be cured of chromatic aberration.

Newton consequently developed a new type of telescope, the reflector, in which there is no objective lens. In this type of telescope, light waves from a distant object enter the open top end and travel down the tube until they hit a mirror at the bottom. This mirror is concave—thicker at the edges than in the middle. Because of this primary mirror's shape, all wavelengths of the light are reflected equally back up the tube to a focus, where a small, flat secondary mirror reflects the image to an eyepiece on the side of the telescope.

Newtonian reflectors were not free of problems. The mirrors, which were made from metal, were hard to grind. The mirror surface also tarnished quickly and had to be polished every few months. These problems kept the Newtonian reflector from being widely accepted until after 1774. At this time, English astronomer William Herschel (1738–1822) developed new designs, polishing techniques, the use of silvered glass, and other innovations that made the reflector telescope much more efficient. In 1781 Herschel discovered the planet Uranus using a reflector telescope he had made. He continued to build reflecting telescopes over the next several years, resulting in the construction of a large telescope in 1789 that housed a mirror with a diameter of almost 4 feet (1.2 meters).

Herschel's telescope remained the largest in the world until 1845, when Irish astronomer William Parsons (1800–1867) constructed a 56-foot-long (17-meter-long) reflector telescope in present-day Birr, Ireland, that came to be known as the Leviathan of Parsonstown. Its mirror, made from speculum metal (an alloy of four parts copper to one part tin), measured 6 feet (1.8 meters) in diameter. Because it tarnished so rapidly in the damp climate, the mirror had to be repolished every six months. Parsons had two mirrors built so that one could be used in the telescope while the other was being repolished. With this telescope, Parsons carried out pioneering astronomical observations, chiefly devoting his time to the study of nebulae (clouds of dust and gas). He was the first to de-

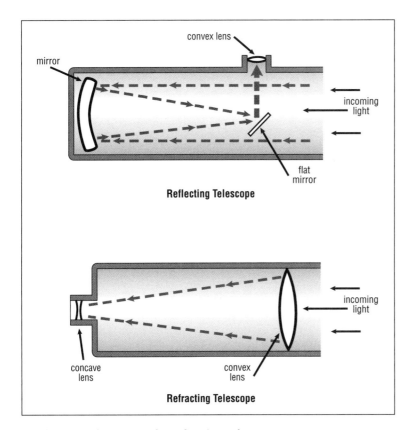

A reflecting telescope and a refracting telescope. *(The Gale Group)*

scribe the spiral nature of nebulae, which were eventually recognized as galaxies outside of the Milky Way.

Modern telescopes and their location

The largest refractor telescopes were built at the end of the nineteenth century and the beginning of the twentieth century. It is easy to understand that larger telescopes are preferred because they gather more light. Astronomical distances are so great that most objects are not visible to the unaided eye. The Andromeda galaxy, generally considered the most distant object that can be seen with the naked eye, is the closest galaxy to Earth outside of the Milky Way. To see far out into space beyond this requires large telescopes.

To accomplish this, astronomers generally prefer reflector telescopes. It is easier to build large mirrors than it is to build large lenses. Telescopes are described by the largest lens or mirror they contain. The largest refractor telescope ever built is the Yerkes 40-inch (102-centimeter) telescope, which is located in southeastern Wisconsin. Refractors are limited to fairly small sizes for two reasons: First, since the light must pass through a lens to be focused, the lens must be supported around its outside edge, not from behind. Large lenses tend to sag and distort in shape because of the effects of gravity. Consequently, the focused image is not as sharp as it should be. Second, because the light passes through the lens, the glass must be entirely free of bubbles or other defects that would distort the image. It is difficult and costly to make large pieces of perfect glass.

Reflector telescopes, on the other hand, make use of mirrors. Since the light is reflected from the front surface, the mirrors can be supported from behind. Therefore, they can be made much larger than the lenses in refractor telescopes. The front surface of a reflector mirror is coated with highly reflective (shiny) aluminum or silver. Since the light in a reflector never passes through the mirror, the glass can contain a few bubbles or other flaws. For these reasons, the largest telescopes in the world are reflectors.

Earth's atmosphere, however, continues to challenge the progress of ground-based astronomy. A problem with the atmosphere is its inherent instability. Even on the clearest of nights, images jiggle and quiver due to atmospheric thermal turbulence caused by irregular air motions (a phenomenon similar to heat waves above a hot road). A way to lessen the impact of the atmosphere is to construct observatories in desolate regions at high attitudes where the atmosphere is thinner (and where the glare of urban artificial light, such as street lights, cannot reach). The best ground-based sites for optical and infrared astronomy are Mauna Kea, a volcano on the island of Hawaii that is 13,797 feet (4,205 meters) high, and the mountain peaks in the desert in northern Chile. Other good sites are in the Canary Islands, a group of seven volcanic islands in the Atlantic Ocean off the northwestern coast of Africa, and in the desert regions in the southwestern United States.

Astronomers look for the following characteristics when they select a site for a ground-based observatory:

- **Clear skies:** The best sites on the planet experience clear skies about 75 percent of the time. Most types of astronomical observations cannot be carried out when clouds are present.

- **Dark skies:** The atmosphere scatters city lights, making it impossible to see faint objects in space. Therefore, the best sites are located far away from large cities. (Even with the naked eye, one can see quite clearly the difference between what can be viewed in the night sky in a city and in a rural area.)

- **High and dry:** Water vapor in Earth's atmosphere absorbs infrared radiation. Fortunately, water vapor is concentrated at low altitudes, and so infrared observatories are best located at high altitudes.

- **Stable air:** Light rays are distorted when they pass through turbulent air, with the result that the image seen through a telescope is distorted and blurred. The most stable air occurs over large bodies of water such as oceans, which have a very uniform temperature. (Changes in temperature cause air masses to rise, if they are heated, or sink, if they are cooled.) As a result, the best sites are located on isolated volcanic peaks in the middle of oceans or in coastal mountain ranges.

Space-based observatories, such as the Hubble Space Telescope, provide images that are clearer and much sharper than those obtained by any ground-based observatory. Astronomers are, however, devising techniques called adaptive optics than can correct atmospheric distortions by changing the shapes of small mirrors hundreds of times each second to compensate precisely for the effects of Earth's atmosphere. Even when this technique is perfected, space-based observatories still will be needed to observe gamma rays, X rays, ultraviolet radiation, and wavelengths of other forms of electromagnetic radiation that are absorbed by the planet's atmosphere before they reach the ground.

Radio astronomy

Viewing requirements for radio observatories are not nearly so rigid as for optical or infrared observatories, and many types of radio observations can be made through clouds.

The largest refractor telescope ever built is the Yerkes forty-inch telescope, which is located in southeastern Wisconsin. *(© Jim Sugar/Corbis)*

Consequently, astronomers in countries that do not have good optical or infrared sites, such as Great Britain, Japan, the Netherlands, and Germany, have concentrated on radio astronomy.

While they may not be bothered much by clouds or city lights, radio telescopes are affected by electrical interference generated by cell phones, radio transmitters, and other electrical devices in present-day society. To counteract this, they are often located far away from large population centers in special radio-quiet zones. Also, certain radio wavelengths are reserved for the use of radio astronomy and cannot be used to transmit human signals. Currently, commercial companies cannot transmit radio waves at frequencies between 71 and 275 gigahertz. (A hertz is a unit used to measure frequency. One hertz equals one cycle or wave per second. One kilohertz [kHz] equals one thousand waves per second, one megahertz

Largest Optical Refracting Telescopes

Observatory	Lens Diameter	Year Built
Yerkes Observatory (Williams Bay, WI)	40 inches (102 centimeters)	1897
Lick Observatory (Mt. Hamilton, CA)	36 inches (91 centimeters)	1888
Paris Observatory (Meudon, France)	32.7 inches (83 centimeters)	1891
Potsdam Observatory (Potsdam, Germany)	31.5 inches (80 centimeters)	1899
Côte d'Azur Observatory (Nice, France)	30 inches (76 centimeters)	1887
Allegheny Observatory (Pittsburgh, PA)	30 inches (76 centimeters)	1914
Royal Greenwich Observatory (London, England)	28 inches (71 centimeters)	1894
Vienna Observatory (Vienna, Austria)	27 inches (69 centimeters)	1878
Berlin Observatory	26.8 inches (68 centimeters)	1896
Johannesburg Observatory	26.4 inches (67 centimeters)	1925
McCormick Observatory (Charlottesville, VA)	26.25 inches (66.7 centimeters)	1883
U.S. Naval Observatory (Washington, DC)	26 inches (66 centimeters)	1873

[MHz] equals one million waves per second, and one gigahertz [GHz] equals one billion waves per second. A typical cell phone operates at 800 MHz. Microwave ovens usually work at a frequency of 2.45 GHz.) The range of radio band frequencies from 30 GHz to 300 GHz is sometimes called the Extremely High Frequency (EHF) range. It corresponds to the millimeter-wave region of the electromagnetic spectrum. Waves in this region have wavelengths from 0.4 inches (10 millimeters) to 0.04 inches (1 millimeter). This means that they are larger than infrared waves or X rays, for example, but smaller than radio waves or microwaves.

No one individual can be given complete credit for the development of radio astronomy. However, an important pioneer in the field was Karl Jansky (1905–1950), a scientist employed at the Bell Telephone Laboratories in Murray Hill, New Jersey. In the early 1930s, Jansky was working on the problem of noise sources that might interfere with the transmission of short-wave radio signals. During his research, Jansky discovered that his instruments picked up static every day at about the same time and in about the same part of the sky. It was later discovered that the source of this static was the center of the Milky Way.

Since then, scientists have found that radio signals come from everywhere. In 1955 astronomers detected radio bursts coming from Jupiter. Next to the Sun, this planet is the strongest source of radio waves in the solar system. Around this time, Dutch astronomer Jan Oort (1900–1992) used a radio telescope to map the spiral structure of the Milky Way. In 1960 several small but intense radio sources were discovered that did not fit into any previously known classification. They were called quasi-stellar radio sources. Further investigation revealed them to be quasars, the most distant and therefore the oldest celestial objects known. And in the late 1960s, English astronomers Antony Hewish (1924–) and Jocelyn Bell Burnell (1943–) detected a strong radio source in the core of the Crab Nebula, a cloud of gas created by a supernova, or the massive explosion of a relatively large star at the end of its lifetime. The source turned out to be the first pulsar ever discovered. (A pulsar is a rapidly spinning, blinking neutron star, which is the extremely dense, compact, neutron-filled remains of a star following a supernova.)

In 1964 radio astronomers found very compelling evidence in support of the big bang theory of how the universe began. U.S. scientists Arno Penzias (1933–) and Robert Wilson (1936–) discovered a constant background noise that seemed to come from every direction in the sky. Further investigation revealed this noise to be radiation (now called cosmic microwave background radiation) that had a temperature of –465°F (–276°C). This corresponded to the predicted temperature to which radiation left over from the formation of the universe about thirteen billion years ago would have cooled by the present.

The largest single radio dish in operation at the beginning of 2004 was that of the Arecibo Observatory, located 10 miles (16 kilometers) south of Arecibo, Puerto Rico. The telescope's 1,000-foot-diameter (305-meter-diameter) spherical reflector consists of almost 40,000 perforated aluminum panels, each

The largest single radio dish in operation at the beginning of 2004 was that of the Arecibo Observatory in Puerto Rico. The telescope's 1,000-foot-diameter spherical reflector consists of almost 40,000 perforated aluminum panels. *(© Bettmann/Corbis)*

---

measuring about 3 feet (1 meter) by 6 feet (2 meters). The panels focus incoming radio waves on movable antenna structures positioned about 450 feet (137 meters) above the reflector surface. It is the largest curved focusing antenna on the planet. Construction of the telescope, which was built in a mountaintop sinkhole (bowl-like depression created when underground rock erodes away), was completed in 1963.

At the end of 2003 construction had begun on the Atacama Large Millimeter Array (ALMA). It is located at one of the driest spots on Earth: a large plateau at an altitude of

16,400 feet (5,000 meters) in the Atacama Desert in northern Chile. The system will consist of 64 radio antennas, each measuring 39 feet (12 meters) in diameter, arranged in an array, with the separations between the antennas varying from 490 feet to 6.2 miles (10 kilometers). When completed, the ALMA will be the largest and most sensitive instrument in the world that measures millimeter and submillimeter wavelengths. The combined antennae will work as an interferometer, a device that uses two or more telescopes to observe the same object at the same time in the same wavelength to increase angular resolution.

Resolution refers to the fineness of detail that can be seen in an image. The larger the telescope, the larger the detail that can be observed. One way to see finer detail is to build a larger single telescope. Unfortunately, there are practical limits to the size of a single telescope. Currently, that is about 33 feet (10 meters) for optical and infrared telescopes and 330 feet (100 meters) for radio telescopes. If, however, astronomers combine the signals from two or more widely separated telescopes, they can see the fineness of detail that would be observed if they had a single telescope of that same diameter. Telescopes working in combination in this way are called interferometers.

Radio interferometry is easier to achieve than optical and infrared interferometry because of the long wavelengths of radio waves. The equipment used to measure radio waves need not be built to the same precision as optical telescopes, and radio waves are not affected as much by turbulence in Earth's atmosphere. For these reasons, radio astronomers have been able to build whole arrays of telescopes separated by thousands of miles to conduct interferometry. For example, U.S. astronomers operate the Very Long Baseline Array, which consists of ten telescopes located across the United States and in the Virgin Islands and Hawaii. When combined with a telescope in Japan, this array of radio telescopes has the same resolution as a telescope with the diameter of Earth.

Infrared astronomy

Infrared astronomy was developed in the 1960s. To be useful, infrared detectors require long periods of time without motion. Since water vapor in the atmosphere is a main interfering substance, infrared astronomy is ideally conducted in space.

But astronomers have used high altitude balloons and ground-based observatories to conduct successful observations in infrared astronomy. Since every object that has a temperature radiates heat energy (infrared radiation), infrared astronomy involves the study of just about everything in the universe.

Infrared telescopes, which are very similar to optical telescopes, have helped astronomers find where new stars are forming, areas known as stellar nurseries. A star forms from a collapsing cloud of gas and dust. Forming and newly formed stars are still covered with dust that blocks optical light. Thus infrared astronomers can more easily probe these stellar nurseries than optical astronomers can. The view of the center of the Milky Way is also blocked by large amounts of interstellar dust. The galactic center is more easily seen by infrared than by optical astronomers.

With the aid of infrared telescopes, astronomers have also located a number of new galaxies, many too far away to be seen through visible light waves. Some of these are dwarf galaxies, which are more plentiful, but contain fewer stars, than visible galaxies. The discovery of these infrared dwarf galaxies has led to the theory that they once dominated the universe and then came together over time to form visible galaxies, such as the Milky Way.

With the growing use of infrared astronomy, astronomers have learned that galaxies contain many more stars than had ever been imagined. Infrared telescopes can detect radiation from relatively cool stars, which give off no visible light. Many of these stars are the size of the Sun. These discoveries have drastically changed astronomers' calculations of the total mass in the universe.

The United Kingdom Infrared Telescope (UKIRT) is the world's largest telescope dedicated solely to infrared astronomy. It is located in Hawaii near the summit of Mauna Kea at an altitude of 13,760 (4,194 meters). The telescope, which began operating in 1979, has a concave primary mirror that measures 12.5 feet (3.8 meters) in diameter.

Optical astronomy

The two most famous historic optical observatories still operational in the United States are the Mount Wilson Ob-

An Old Telescope Finds a New Icy World

In November 2003, a group of astronomers using a 48-inch (122-centimeter) telescope at the Palomar Observatory in southern California sighted what they believed was the most distant object known to orbit the Sun and the largest one to be detected since the discovery of the planet Pluto in 1930.

The astronomers proposed naming the object Sedna, after the Inuit goddess who created the sea creatures of the Arctic. The object, properly referred to as a planetoid (a very small planet), is estimated to have a diameter of no more than 1,100 miles (1,770 kilometers), roughly 300 miles (430 kilometers) less than that of Pluto. Sedna is also extremely frigid. Temperatures on Sedna are estimated to hover at –400°F (–240°C). Calculations indicate that on its widely eccentric orbit, Sedna takes 10,500 years to travel around the Sun.

A peculiar trait of Sedna is its red color. In the solar system, only Mars matches its color. Astronomers are unsure why it appears red. They are also unsure of its composition, believing that it might be a mix of rock and ice.

servatory and the Palomar Observatory. These two southern California astronomical research centers have an intertwined past, highlighted by legendary astronomers and landmark discoveries. The Mount Wilson Observatory came first, and of the two, it is considered to be the premier observatory.

Located at an altitude of 5,700 feet (1,737 meters) on Mount Wilson, a peak in the San Gabriel Mountains near Pasadena, California, the observatory houses two reflecting telescopes. The first, measuring 60 inches (152 centimeter) in diameter, was installed in 1908. The second, named the Hooker Telescope after John D. Hooker, a local businessman who funded its construction, was installed in 1917. Measuring 100 inches (254 centimeters) in diameter, it remained the largest telescope in the world until 1948.

U.S. astronomer George Hale (1868–1938) had solicited money for the construction of the observatory, which originally was to have been a research center designed specifically for the study of the Sun. As director of the observatory, Hale was also the brainchild behind the Hooker Telescope. During his career, he was the driving force behind the creation of four telescopes, each surpassing the last as the world's largest.

Hubble's great discoveries

The greatest discoveries made with the Hooker Telescope actually were not made by Hale. Instead, it was U.S. astronomer Edwin Hubble (1889–1953) who used the Hooker Telescope to make observations and discoveries that profoundly changed humans' concept of the universe and their place in it. Hubble's first notable achievement at Mount Wilson was the confirmation of the existence of galaxies outside of the Milky Way. From observations he made in 1923, Hubble was able to identify a type of variable star known as a Cepheid (pronounced SEE-fee-id) in a nebula in a region of space known as "Andromeda" (known today as the Andromeda galaxy). Variable stars are so-named because their light output changes over time, varying between dim and bright. By using information about the relationship between brightness, luminosity (how much light a star radiates), and the distances of Cepheid stars in the Milky Way, Hubble was able to estimate the distance to the Cepheid in the nebula to be about one million light-years. (The term light-year refers to the distance light travels in space in one year, about 6 trillion miles [9.6 trillion kilometers]).

Hubble also discovered other Cepheids, as well as other objects, and calculated the distances to them. Since scientists knew that the maximum diameter of the Milky Way was only one hundred thousand light-years, Hubble's figures established the existence of galaxies outside of the Milky Way. Eventually he discovered nine new galaxies.

Continuing his pioneering work on galaxies throughout the 1920s, Hubble determined distances for more than twenty galaxies surrounding the Milky Way. In 1929 this work led to his most important discovery. For more than a decade, scientists had predicted that the light coming from some distant galaxies might indicate that the galaxies were moving apart from each other and Earth. If the galaxies were speeding fast enough away from Earth, the motion would "stretch" the light waves emitted from them. Since longer wavelengths make light take on a reddish tone, this stretching was called the redshift.

Hubble's greatest achievement was to determine the redshifts for a large number of galaxies by measuring the wavelengths of the light coming from them. His measurements led him to two important conclusions. First, distant galaxies did seem to be moving away from Earth. Second, the farther away they were from Earth, the faster they seemed to move. This relationship between a galaxy's distance and its speed is now known as Hubble's law.

Largest Optical Telescopes

Telescope Name	Location	Size of Primary Optical Surface
Keck I	Mauna Kea, Hawaii	33 feet (10 meters)
Keck II	Mauna Kea, Hawaii	33 feet (10 meters)
Hobby-Eberly Telescope	Mount Locke, Texas	30.2 feet (9.2 meters)
Large Binocular Telescope (2 telescopes)	Mount Graham, Arizona	27.6 feet (8.4 meters)
Very Large Telescope (4 telescopes)	Cerro Paranal, Chile	27 feet (8.2 meters)
Subaru Telescope	Mauna Kea, Hawaii	27 feet (8.2 meters)
Gemini North Telescope	Mauna Kea, Hawaii	26.6 feet (8.1 meters)
Gemini South Telescope	Cerro Pachon, Chile	26.6 feet (8.1 meters)
Multiple Mirror Telescope	Mount Hopkins, Arizona	21.3 feet (6.5 meters)
Magellan I	Los Campanas, Chile	21.3 feet (6.5 meters)
Magellan II	Los Campanas, Chile	21.3 feet (6.5 meters)
Bolshoi Teleskop Azimutalnyi ("Large Altazimuth Telescope")	Mount Pastukhov, Russia	19.7 feet (6 meters)

The information collected by Hubble at Mount Wilson supported the big bang theory of the creation of the universe. The movement of galaxies away from one another is consistent with the idea that the universe began as a single point billions of years ago and that a huge explosion resulted in matter being created and scattered over great distances.

Two decades after Hubble developed his famous equations, George Hale built yet another large reflecting telescope, one that held the distinction of being the largest optical telescope in the world for three decades. The 200-inch-diameter (508-centimeter-diameter) Hale Telescope is housed at the Palomar Observatory, which is located at an altitude of 6,000 feet (1,830 meters) on Palomar Mountain 90 miles (145 kilometers) southeast of Mount Wilson Observatory. The observatory currently contains three other instruments, the largest of which is a 60-inch-diameter (152-centimeter-diameter) reflecting telescope.

The 20-ton (18-metric ton) mirror for the Hale Telescope was brought to the mountain and erected inside the twelve-story, 1,000-ton (907-metric ton) rotating dome that had been built specifically for the telescope. Scientific research finally began at the observatory in 1948. In the observatory's early days, German astronomer Walter Baade (1893–1960) identified more than three hundred Cepheid variables in the Andromeda galaxy. And Swiss astronomer Fritz Zwicky (1898-1974), who worked at both Palomar and Mount Wilson observatories, made detailed studies of supernovas, neutron stars, and dark matter (virtually undetectable matter that does not emit or reflect light and that is thought to account for 90 percent of the mass of the universe, acting as a "cosmic glue" that holds together galaxies and clusters of galaxies).

Present-day giants

By the early twenty-first century, more than one dozen mirrors with diameters larger than 21.3 feet (6.5 meters) had been installed in optical and infrared telescopes around the world. The largest single-mirror reflecting optical telescope is the 27-foot (8.2-meter) Subaru Telescope, formerly called the Japanese National Large Telescope, located at the summit of Mauna Kea, Hawaii. Located nearby is the Gemini North Telescope, another large single-mirror reflecting telescope. It measures 26.6 feet (8.1 meters) in diameter. The surface of the mirror is so smooth that if it were enlarged to the size of Earth, the largest bump on the mirror would be only 1 foot (0.3 meter) high. Its twin, the Gemini South Telescope, is located at Cerro Pachon, Chile.

Currently, the largest optical/infrared reflecting telescopes in the world are the twin 33-foot (10-meter) Keck telescopes on Mauna Kea. These telescopes do not contain a single mirror that is 33 feet (10 meters) in diameter. It would be very difficult to manufacture a single mirror that size that would not distort under gravity and become useless. Instead, each mirror consists of thirty-six hexagonal-shaped mirrors that are

The largest optical/infrared reflecting telescopes in the world are the twin 33-foot Keck telescopes on Mauna Kea, Hawaii. Each telescope has a mirror that consists of thirty-six hexagonal-shaped mirrors that fit together like bathroom tiles. *(© Roger Ressmeyer/Corbis)*

6 feet (1.8 meters) in diameter and fitted together like bathroom tiles. A perfect fit between the tiles is ensured by a computer-activated system, which pushes on the back of each segment to counteract the pull of gravity in order to maintain a perfect reflecting shape. The two telescopes, located 279 feet (85 meters) apart, have been operated as an interferometer, mimicking a telescope that has a diameter of 279 feet (85 meters).

The Hobby-Eberly Telescope at the McDonald Observatory, located on top of Mount Locke in the Davis Mountains in western Texas, has a primary mirror that is actually bigger than those of the twin Kecks. It measures 36.4 by 32.2 feet (11.1 by 9.8 meters). However, at any given time during observations, only a 30.2-foot-diameter (9.2-meter-diameter) section of the mirror is utilized, making it the world's third largest optical telescope.

These U.S. telescopes are rivaled in power by the Very Large Telescope (VLT), operated by the European Southern Observatory, an international astronomical organization composed of ten European countries. VLT is located at the Paranal Observatory on the summit of Cerro Paranal, an 8,645-foot (2,635-meter) mountain in the Atacama Desert in northern Chile. VLT consists of a cluster of four telescopes, each containing a mirror almost 27 feet (8.2 meters) in diameter. In the language of the Mapuche, the indigenous people who live in the area, the four telescopes have been given names of astronomical objects: Antu ("Sun"), Kueyen ("Moon"), Melipal ("Southern Cross"), and Yepun ("Venus"). The VLT can be operated as a set of four independent telescopes or as an interferometer. In this latter mode, the VLT mimics a telescope that has a mirror 656 feet (200 meters) in diameter, making it the largest optical telescope in the world.

Astronomers in Europe are exploring the possibility of building an optical/infrared telescope that is 330 feet (100 meters) in diameter. The telescope is called the OWL, which stands for the Overwhelmingly Large Telescope. The mirror would be built in the same way as the Keck mirrors: combining thousands of smaller mirrors to form a single continuous surface.

Because of the vast distances in space, light that travels through the universe also travels through time. Light from an object located five million light-years away from Earth left that object five million years ago. Looking at light in the sky is looking backward in time. The farther one looks out into space, the further one sees back in time. If built, the OWL would be powerful enough to study objects present when the universe was only a few million years old. This would allow astronomers to observe directly the evolution of the universe throughout nearly all of its history.

For More Information

Books

Christianson, Gale E. *Edwin Hubble: Mariner of the Nebulae.* Chicago, IL: University of Chicago Press, 1996.

Florence, Ronald. *The Perfect Machine: Building the Palomar Telescope.* New York: HarperCollins, 1994.

Orr, Tamra B. *The Telescope.* New York: Franklin Watts, 2004.

Panek, Richard. *Seeing and Believing: How the Telescope Opened Our Eyes and Minds to the Heavens.* New York: Penguin, 1999.

Parker, Barry R. *Stairway to the Stars: The Story of the World's Largest Observatory.* New York: Perseus Publishing, 2001.

Web Sites

Mount Wilson Observatory. http://www.mtwilson.edu/ (accessed on August 19, 2004).

NASA Space Technology Ground-based Solar and Astrophysical Observatory Guide. http://ranier.oact.hq.nasa.gov/Sensors_page/GroundObserv. html (accessed on August 19, 2004).

National Radio Astronomy Observatory. http://www.nrao.edu/ (accessed on August 19, 2004).

"Paranal Observatory." *European Southern Observatory.* http://www.eso. org/paranal/ (accessed on August 19, 2004).

W. M. Keck Observatory. http://www2.keck.hawaii.edu/ (accessed on August 19, 2004).

13

Space-based Observatories

Astronomy is the scientific study of the physical universe beyond Earth's atmosphere. At its most basic, astronomy is essentially about the observation of light. The ability to gather light is the key to acquiring useful astronomical information. The larger the primary mirror of a telescope, for example, the greater its light-gathering capabilities and the greater the magnification of the instrument. These two attributes allow a large telescope to view fainter, smaller objects than a telescope of lesser size.

Astronomy is not just about visible light, however. Visible light, also known as optical radiation, is only one form of electromagnetic radiation, a collective term for radiation consisting of electric and magnetic waves that travel through space at the speed of light, approximately 186,000 miles (299,274 kilometers) per second. Like any wave, those that make up the forms of electromagnetic radiation can be described by two properties: wavelength and frequency. The wavelength is the distance between two successive identical parts of the wave, such as between two wave peaks or crests. Frequency is the rate at which two successive identical parts

Launched in 1990, the Hubble Space Telescope was the first of four "Great Observatories" studying the universe from space.

(National Aeronautics and Space Administration)

of the wave pass a given point. Wavelength and frequency have a reciprocal relationship with each other: As one increases, the other must decrease.

The various forms of electromagnetic radiation make up the electromagnetic spectrum much as the various colors of light make up the visible spectrum (the rainbow). The visible portion of the electromagnetic spectrum is most familiar because human eyes are optimized for these wavelengths. Observations made in the visible region show only a small portion of the activities and processes underway in the universe. When astronomers view the sky in the invisible regions of the electromagnetic spectrum, they see an entirely different picture.

The different forms of radiation in the electromagnetic spectrum, in order from lowest to highest energy, are: radio, microwaves, infrared, optical or visible, ultraviolet, X rays, and gamma rays.

Radio waves have the lowest frequency and longest wavelength of any form of electromagnetic radiation. Their wavelengths measure up to 6 miles (9.7 kilometers) from peak to peak. Radio stations on Earth transmit information (music or speech) through these energy waves. In space, many celestial objects emit radio waves, including the Sun, quasars (pronounced KWAY-zarz; extremely bright objects found in remote areas of space), pulsars (rapidly rotating neutron or burned-out stars), gas clouds, and the centers of galaxies.

Microwaves have wavelengths between 0.04 and 11.8 inches (0.1 and 30 centimeters). Their short wavelengths make them ideal for use in radio and television broadcasting. They are also used to cook food and to communicate with artificial satellites (man-made devices that orbit Earth and other celestial bodies). Astronomers use microwaves to study the universe. Microwave radiation is the firmest evidence in support of the big bang theory, which explains the beginning of the universe as a tremendous explosion from a single point that occurred about thirteen billion years ago. In the 1960s, radio astronomers discovered microwaves coming from every direction in space. This radiation, now called cosmic microwave background radiation, is believed to be the radiation left over from the big bang.

Words to Know

Antimatter: Matter that is exactly the same as normal matter, but with the opposite spin and electrical charge.

Artificial satellite: A man-made device that orbits Earth and other celestial bodies and that follows the same gravitational laws that govern the orbit of a natural satellite.

Astronomy: The scientific study of the physical universe beyond Earth's atmosphere.

Big bang theory: The theory that explains the beginning of the universe as a tremendous explosion from a single point that occurred about thirteen billion years ago.

Binary star: A pair of stars orbiting around one another, linked by gravity.

Black hole: The remains of a massive star that has burned out its nuclear fuel and collapsed under tremendous gravitational force into a single point of infinite mass and gravity from which nothing escapes, not even light.

Brown dwarf: A small, cool, dark ball of matter that never completes the process of becoming a star.

Corona: The outermost and hottest layer of the Sun's atmosphere that extends out into space for millions of miles.

Dark matter: Virtually undetectable matter that does not emit or reflect light and that is thought to account for 90 percent of the mass of the universe, acting as a "cosmic glue" that holds together galaxies and clusters of galaxies.

Electromagnetic radiation: Radiation that transmits energy through the interaction of electricity and magnetism.

Electromagnetic spectrum: The entire range of wavelengths of electromagnetic radiation.

Gamma rays: Short-wavelength, high-energy radiation formed either by the decay of radioactive elements or by nuclear reactions.

Geosynchronous orbit: An orbit in which a satellite revolves around Earth at the same rate at which Earth rotates on its axis; thus, the satellite remains positioned over the same location on Earth.

Inflationary theory: The theory that the universe underwent a period of rapid expansion immediately following the big bang.

Infrared radiation: Electromagnetic radiation with wavelengths slightly longer than those of visible light.

Interstellar: Between or among the stars.

Interstellar medium: The gas and dust that exists in the space between stars.

Light-year: The distance light travels in the near vacuum of space in one year, about 5.88 trillion miles (9.46 trillion kilometers).

Microwaves: Electromagnetic radiation with a wavelength longer than infrared radiation but shorter than radio waves.

Neutron star: The extremely dense, compact, neutron-filled remains of a star following a supernova.

Observatory: A structure designed and equipped to observe astronomical phenomena.

Ozone layer: An atmospheric layer that contains a high proportion of ozone molecules that absorb incoming ultraviolet radiation.

Pulsar: A rapidly spinning, blinking neutron star.

Quasars: Extremely bright, star-like sources of radio waves that are found in remote areas of space and that are the oldest known objects in the universe.

Radiation: The emission and movement of waves of atomic particles through space or other media.

Radio waves: The longest form of electromagnetic radiation, measuring up to 6 miles (9.7 kilometers) from peak to peak in the wave.

Redshift: The shift of an object's light spectrum toward the red end of the visible light range, which is an indication that the object is moving away from the observer.

Reflector telescope: A telescope that directs light from an opening at one end to a concave mirror at the far end, which reflects the light back to smaller mirror that directs it to an eyepiece on the side of the telescope.

Solar flare: A temporary bright spot that explodes on the Sun's surface, releasing an incredible amount of energy.

Solar wind: Electrically charged subatomic particles that flow out from the Sun.

Spacewalk: Technically known as an EVA, or extravehicular activity, an excursion outside a spacecraft or space station by an astronaut or cosmonaut wearing only a pressurized spacesuit and, possibly, some sort of maneuvering device.

Spectrograph: A device that separates light by wavelengths to produce a spectrum.

Stellar wind: Electrically charged subatomic particles that flow out from a star (like the solar wind, but from a star other than the Sun).

Sunspot: A cool area of magnetic disturbance that forms a dark blemish on the surface of the Sun.

Supernova: The massive explosion of a relatively large star at the end of its lifetime.

Ultraviolet radiation: Electromagnetic radiation of a wavelength just shorter than the violet (shortest wavelength) end of the visible light spectrum.

Van Allen belts: Two doughnut-shaped belts of high-energy charged particles trapped in Earth's magnetic field.

X rays: Electromagnetic radiation of a wavelength just shorter than ultraviolet radiation but longer than gamma rays that can penetrate solids and produce an electrical charge in gases.

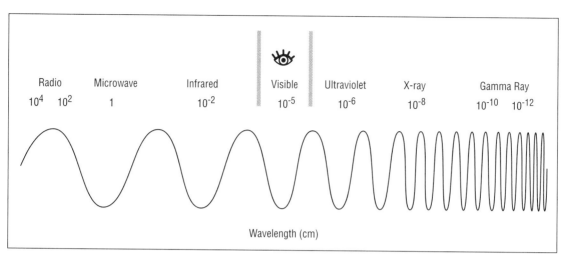

Radio	Microwave	Infrared	Visible	Ultraviolet	X-ray	Gamma Ray
10^4 10^2 1		10^{-2}	10^{-5}	10^{-6}	10^{-8}	10^{-10} 10^{-12}

Wavelength (cm)

The electromagnetic spectrum. *(The Gale Group)*

The wavelengths of infrared radiation, which measures from 0.00003 to 0.04 inch (0.000075 to 0.1 centimeter), are slightly longer than those of visible light. Humans cannot see infrared radiation, but can sense its energy as heat on the skin. Infrared radiation is emitted by any object that has a temperature (radiates heat). So, basically all celestial objects emit some infrared radiation.

Optical radiation is visible light. The different colors of light the human eye can see correspond to different wavelengths: Red light has the longest wavelength, violet the shortest. Moving from red to violet are the remaining colors: orange, yellow, green, blue, and indigo. Optical radiation is emitted by everything from fireflies to light bulbs to stars.

Ultraviolet (UV) radiation has wavelengths just shorter than the violet end of the visible light spectrum. The Sun at the center of our solar system is a major source of UV radiation. Too much UV radiation is harmful to living organisms. It burns skin, leads to the development of skin cancer, and damages vegetation. Fortunately, the ozone layer in Earth's atmosphere prevents most UV radiation from reaching the planet's surface. Stars and other hot celestial objects emit UV radiation as well.

X rays, which have wavelengths just shorter than ultraviolet radiation but longer than gamma rays, can penetrate solids and produce an electrical charge in gases. Earth's atmosphere filters out most X rays, which in a large dose would be deadly to humans and other forms of life on the planet. X-ray astronomy is a relatively new scientific field focusing on celestial objects that emit X rays. Such objects include stars, galaxies, quasars, pulsars, and black holes (the remains of massive stars that have burned out their nuclear fuel and collapsed under tremendous gravitational force into single points of infinite mass and gravity from which nothing escapes, not even light).

Gamma rays are short-wavelength, high-energy radiation formed either by the decay of radioactive elements or by nuclear reactions. Gamma rays produced on Earth are known as terrestrial gamma rays. They are the only gamma rays that can be observed on the planet. Those gamma rays produced in space, cosmic gamma rays, do not penetrate to the surface of Earth because the atmosphere absorbs this high-energy radiation. Gamma rays in space are created by highly energetic reactions. Only the hottest, most active celestial objects give off gamma rays: solar flares (powerful eruptions on the surface of the Sun), supernova explosions (massive explosions of relatively large stars at the end of their lifetime), neutron stars, pulsars, and black holes, among others.

Seeing beyond Earth

For life-forms on Earth, it is fortunate that many of these forms of electromagnetic radiation do not reach the planet's surface. In the field of astronomy, though, this creates a problem. In order to study the universe as fully as possible, astronomers have been forced to place observatories beyond Earth, either in orbit around the planet or in deep space. Space-based observatories, however, are typically more complicated and more expensive than ground-based observatories. The National Aeronautics and Space Administration (NASA) and other space agencies have placed observatories in space since the latter part of the 1960s. While the Hubble Space Telescope is perhaps the most famous of the space-based observatories, it is just one of many that have provided astronomers with new insights about our solar system, the Milky Way (the galaxy that contains our solar system), and the universe.

Observatories in space have a number of key advantages. Telescopes in space are able to operate twenty-four hours a day, free from both Earth's day-night cycle as well as clouds and other weather conditions that can interfere with observing. They are also subject to neither light pollution from artificial light sources on Earth nor the heat distortions in the atmosphere that blur images when viewed from the ground. Telescopes above the atmosphere can observe those portions of the electromagnetic spectrum, such as UV, X rays, and gamma rays, that are blocked by Earth's atmosphere and never reach the surface. And they are better able to view infrared radiation, which is partially blocked by the atmosphere. Because of this, space-based observatories are more productive and useful than their ground-based counterparts.

There are some disadvantages to space-based observatories, however. Unlike most telescopes on the ground, those in space operate completely automatically, without any humans on-site to fix faulty equipment or handle any other problems that arise. There are also limitations on the size and mass of objects that can be launched. Special materials and designs that can withstand the harsh environment of space must be utilized, limiting the types of observatories that can be placed in space. These factors, as well as high launch costs, make space-based observatories very expensive: The largest observatories in space, such as the Hubble Space Telescope, cost more than one billion dollars; the best observatories on the ground cost less than one hundred million dollars. Despite the great cost, there is no question that space-based observatories are crucial to gathering the information needed to help humans understand the universe.

The beginning of telescopes in space

The first serious study of observatories in space was conducted in 1946 by U.S. astrophysicist Lyman Spitzer (1914–1997; an astrophysicist is an astronomer who studies the physical properties of celestial bodies). More than a decade before the launch of the first artificial satellite into space and twelve years before the formation of NASA, Spitzer wrote a paper describing in detail the advantages of putting a telescope in space. He believed that observations made in space could revolutionize the field of astronomy. A space-based observatory, he maintained, would be able to detect a wide range of electromagnetic wavelengths and not be hindered by the blurring effects of the atmosphere. He also thought that a telescope in space would reveal much clearer images, of even far-off objects, than any ground-based telescope.

Some twenty years would pass before the first space-based observatories would be launched into orbit. Among the first was a series of astronomical satellites that NASA launched under the name Orbiting Astronomical Observatories (OAO). The purpose of these satellites was to provide astronomical data about UV radiation and X rays. OAO-1 was launched successfully on April 8, 1966, from Cape Canaveral, Florida, but its primary battery overheated two days later and it stopped working.

More than two years later, on December 7, 1968, NASA launched OAO-2. The second satellite in the series proved highly successful. It carried eleven telescopes, and at 4,432 pounds (2,012 kilograms), it was the heaviest satellite placed in orbit up to that time.

The first serious study of observatories in space was conducted in 1946 by U.S. astrophysicist Lyman Spitzer. *(AP/Wide World Photos)*

Over a period of more than four years, it made 22,560 X-ray, UV, and infrared observations of stars. In May 1972, its instruments detected a supernova. OAO-2 was also the first space-based observatory to detect UV radiation coming from the center of the Andromeda galaxy, the nearest galaxy to the Milky Way and the most distant object that can be seen with the naked eye in the night sky.

OAO-B, the replacement for OAO-1, was also lost. Launched on November 30, 1970, it failed to achieve Earth orbit and fell into the Atlantic Ocean. The series was saved, however, by OAO-3, perhaps the most successful of these early astronomical satellites. Launched on August 21, 1972, it was later renamed *Copernicus* in honor of the five hundredth anniversary of the birth of Polish astronomer Nicolaus Copernicus (1473–1543). Until early 1981, it returned data on the birth, death, and life cycles of stars. OAO-3 was a collaborative effort between NASA and the Science and Engineering Research Council of the United Kingdom. The main experiment onboard was the 990-pound (450-kilogram) Princeton Experiments Package that included a 31.5-inch (80-centimeter) UV telescope, the largest telescope used in space up to that time. The primary purpose of OAO-3 was to study UV radiation from stars near the edges of the Milky Way and from gas and dust in interstellar space, the space between the stars. (Some sources state that interstellar space is space starting from the edge of the solar system and extending to the limit of the Milky Way. They call the area beyond this intergalactic space.) OAO-3 also studied X rays from stars and X-ray absorption in interstellar space.

The most productive astronomical telescope

NASA followed up the OAO series with a number of other small observatories. Then, on January 26, 1978, NASA launched the most successful UV satellite, and perhaps the most productive astronomical telescope ever. The International Ultraviolet Explorer (IUE) was put into a geosynchronous orbit 32,475 miles (52,250 kilometers) over the Atlantic Ocean, the first space-based observatory to be placed in such a high orbit. (A geosynchronous orbit is one in which a satellite travels around Earth in the same time it takes the planet to rotate once on its axis. Thus, the satellite always remains in the same position relative to Earth's surface.) Weighing 1,480 pounds (672 kilograms), the IUE measured 14 feet by 5 feet by 5 feet (4.3 meters by 1.5 meters by 1.5 meters) and was powered by two solar panels. The IUE was equipped with a 17.7-inch (45-centimeter) telescope hooked up with two spectrographs, devices that separate light by wavelength, allowing for the identification of elements within the light source. The spectrographs recorded UV wavelengths and transmitted the information back to observatories on Earth. At the time, the IUE was the only satellite observatory that worked continually twenty-four hours a day.

The IUE, which was a joint project of NASA, PPARC, and the European Space Agency (ESA), was intended to function for three to five years. However, it lasted almost nineteen years. During its lengthy service to the astronomy community, it did suffer some minor mechanical problems. Although

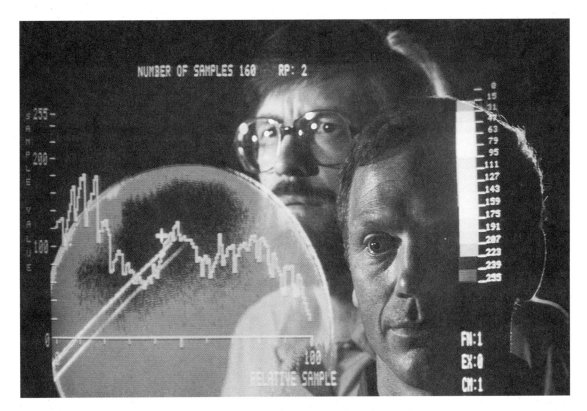

NUMBER OF SAMPLES 160   RP: 2

Astronomers George Sonneborn and Robert Kirschner study data that was gathered by the International Ultraviolet Explorer (IUE). The satellite was launched in 1978 and is considered the most productive astronomical telescope ever. *(© Roger Ressmeyer/Corbis)*

engineers were able to keep the IUE functioning at various capacity levels, the final shutdown occurred on September 30, 1996, after a joint decision by NASA and ESA.

Built to explore astronomical objects such as stars, comets, galaxies, and supernovae (plural of supernova) that exist in the UV portion of space, the IUE made observations of more than one hundred thousand astronomical objects during its use. Scientists from all over the world have used the information that the IUE was able to collect. More than thirty-five hundred scientific articles have been generated from this information, which is the most productive for any observatory satellite to date.

The IUE made history when it helped make the first identification of an exploding star, named Supernova 1987A, in

February 1987. In March 1996 the IUE observed the nucleus of the newly discovered Comet Hyakutake (pronounced hyah-koo-TAH-key) as it underwent chemical changes during its five-day breakup. The telescope continually sent pictures to Earth, and from them scientists learned that every time the comet passed the Sun, it ejected ten tons (nine metric tons) of water every second and that the eventual breakup of the comet involved only a very small piece of the comet. Other major milestones of the IUE included studies of stellar winds (charged particles ejected from a star's surface), hot gas around the Milky Way, and stars with magnetic fields and surface activity.

The IUE was succeeded by the Extreme Ultraviolet Explorer (EUVE), which was launched on June 7, 1992. The EUVE was designed to extend the UV spectral coverage of the IUE by being able to observe much shorter wavelengths. (Extreme ultraviolet light is located in the spectrum between ultraviolet light and X rays.) It contained four telescopes: three scanning survey telescopes and one deep survey/spectrometer telescope. The metal mirrors in each one measured 15.7 inches (40 centimeters) in diameter. Each scanning survey telescope, about as large as a 55-gallon (25-kilogram) oil drum, weighed about 260 pounds (118 kilograms). The deep survey telescope/spectrometer weighed about 710 pounds (322 kilograms). The EUVE had a total mass of 7,215 pounds (3,275 kilograms).

Since little was known about extreme ultraviolet radiation, astronomers hoped to learn, through EUVE observations, about the physical properties and chemical compositions of stars, planets, and other sources of extreme ultraviolet radiation. During the first six months of its operation, the EUVE conducted the first extreme ultraviolet survey of the sky. It then began making pointed observations, mainly with its deep survey telescope/spectrometer. By the time science operations on the telescope ended on January 26, 2001, it had collected data from more than one thousand objects in the Milky Way and more than thirty-six that lie beyond. A little more than a year later, on January 30, 2002, the EUVE was allowed to reenter Earth's atmosphere where it broke apart.

A third major UV satellite, the Far Ultraviolet Spectroscopic Explorer (FUSE), was launched on June 24, 1999. It, too, was designed to look farther into the UV portion of the electromagnetic spectrum, observing those wavelengths with much greater sensitivity and resolving power than previous instruments. It carried four 13.8-inch-diameter (35-centimeter-diameter) far UV telescopes, each with a UV high-resolution spectrograph. By observing the far UV light from stars, interstellar gas, and distant galaxies with FUSE, astronomers had hoped to understand the properties of interstellar gas clouds, how chemical elements are dispersed throughout galaxies, and, perhaps most important, what conditions were like in the universe during the first few minutes after the big bang (the theory that explains the beginning of the universe as a tremendous explosion from a single point that occurred about thirteen billion years ago). The primary mission of FUSE was scheduled to last three-and-one-half years. By the end of that time, it had recorded more than eight thousand hours of data. The mission of the satellite was then extended, and as of 2004, it continued to provide astronomers with observations.

COBE and the big bang

The search for the beginning of the universe began, in earnest, in the twentieth century. Before then, astronomers and others assumed that the universe had always existed as it was, without any changes. In the 1920s, however, U.S. astronomer Edwin Hubble (1899–1953) discovered observable proof that other galaxies existed in the universe besides the Milky Way. In 1929 he made his most important discovery: All matter in the universe was moving away from all other matter in all directions. This proved that the universe was expanding. Hubble also discovered that galaxies located farther away from Earth seemed to be moving away at a faster rate.

Based on Hubble's discoveries, cosmologists (scientists who study the origin of the universe) developed the big bang theory. By the mid-1960s it had become the foremost scientific model used to describe the creation of the universe. However, some problems with the theory still remained. When the big bang occurred, hot radiation given off by the explosion expanded and cooled with the universe. Known as the cosmic microwave background radiation, this radiation appears as a weak hiss of radio noise coming from all directions in space. It is, in a sense, the oldest light in the universe. When astronomers measured this radiation, they found its temperature to be just less than –450°F (–270°C). Scientists estimate that this would be the approximate temperature of the universe if it had expanded and cooled since the big bang.

The radiation, though, seemed smooth, with no temperature fluctuations. If the universe had cooled at a steady rate, it would have had to expand and cool at a steady rate. If this were true, planets and galaxies would not have been able to form. Gravity, which would help them clump together, would have caused fluctuations in the temperature readings. In 1980 U.S. physicist and cosmologist Alan Guth (1945–) proposed a supplemental idea to the big bang theory. Called the inflationary theory, it suggests that at first the universe expanded at a much faster rate than it does now. This concept of accelerated expansion allows for the formation of the stars and planets.

To measure the cosmic microwave background radiation, NASA launched the Cosmic Background Explorer (COBE) on November 18, 1989. COBE carried instruments that searched for the cosmic microwave background radiation and precisely mapped it. Guth's inflationary theory was supported in April 1992 when NASA announced that COBE had detected tiny temperature fluctuations in the background. Scientists regard these fluctuations as proof that gravitational disturbances existed in the early universe, which allowed matter to clump together to form large stellar bodies such as galaxies and planets. NASA ended COBE operations in December 1993. Although it no longer returns scientific data, COBE remains in orbit.

Present-day astronomers liken the study of cosmic microwave background radiation in cosmology to that of DNA (deoxyribonucleic acid; the complex molecule that stores and transmits genetic information) in biology. They consider it the seed from which stars and galaxies grew. To widen the scope and precision of that study, NASA launched a satellite called the Wilkinson Microwave Anisotropy Probe (WMAP) on June 30, 2001. Orbiting much farther away from Earth than COBE, 931,500 miles (1,500,000 kilometers), the goal of the WMAP was to measure temperature differences in the cosmic microwave background on a much finer scale. Astronomers had hoped that the information gathered by the WMAP would reveal a great deal about the universe, including its shape as it appeared about thirteen billion years ago.

Hubble: The first of the Great Observatories

While NASA had been developing and launching its early space-based observatories, it had also been working on something much larger. In the 1960s it started studying a proposal to launch a large observatory that would study optical, UV, and a portion of near-infrared (the "red end" of the optical spectrum) wavelengths. This observatory was originally known simply as the Large Space Telescope, but was renamed the Hubble Space Telescope (HST) after Edwin Hubble in the 1970s. NASA planned for Hubble to be the first of four "Great Observatories" studying the universe from space, each focusing on a different portion of the electromagnetic spectrum.

In 1977 ESA joined NASA as a partner in the project, with an agreement to supply 15 percent of the equipment needed for the HST in exchange for 15 percent of the observing time. In 1985, after eight years of construction, the 2.1-billion-dollar HST was finally ready for launch. But then, in January 1986, came the explosion of the space shuttle *Challenger,* an accident that led to the grounding of the entire shuttle fleet for the next two years and eight months. The HST launch was delayed until April 24, 1990, when it was granted a spot on the space shuttle *Discovery*.

Weighing 24,255 pounds (11,000 kilograms), the HST is 43.3 feet (13.2 meters) long and has a maximum diameter of 13.8 feet (4.2 meters). It is a reflector telescope, a type of telescope that directs light from an opening at one end to a concave mirror at the far end. The concave mirror reflects the light back to a smaller mirror that directs it to an eyepiece on the side of the telescope. The HST is equipped with two mirrors; the main high-quality mirror, which has a diameter of about 7.9 feet (2.4 meters), can detect a lighted candle more than 250,000 miles (402,250 kilometers) away. The light collected and focused by the telescope ends up in one of four instruments: three cameras and a spectrograph. The HST also carries computers that can receive commands from its data-gathering site, the Space Telescope Institute in Baltimore, Maryland. Two solar panels provide electricity, which is used mainly to power the cameras and the four large gyroscopes used to orient and stabilize the telescope. (A gyroscope is an instrument consisting of a frame supporting a disk or wheel that spins rapidly about an axis. Once a gyroscope is set spinning, no amount of tilting or turning will change the direction

Deepest Glance Into the Universe

At a time when the future of the Hubble Space Telescope (HST) looked bleak, the space-based observatory captured an image that offered astronomers a rare glimpse back to when the universe was just 750 million years old.

Between September 24, 2003, and January 16, 2004, the HST made a one-million-second-long exposure of a small region of space now called the Hubble Ultra Deep Field. In the image, captured over the course of four hundred Hubble orbits around Earth, are an estimated ten thousand galaxies. It is the deepest image of the universe ever taken in visible light, looking back in time more than thirteen billion years. Astronomers believe the galaxies may be the oldest and most distant known to date. Astronomers can estimate how old a galaxy is by measuring the light it emits, specifically that amount of light that has been shifted toward the red end of the visible spectrum. The higher the redshift of the galaxy, the more distant it is and the earlier it existed in the universe.

Astronomers call the time when these ancient galaxies emerged the "Dark Ages" of the universe. Further images from this period in the history of the universe will not be possible until at least sometime between 2009 and 2011, when the James Webb Space Telescope is scheduled for launch.

in which it is pointing.) The HST also contains batteries that power certain systems. It orbits Earth at a distance of 373 miles (600 kilometers), completing one revolution around the planet every 100 minutes.

Two months after the HST had been placed in orbit, NASA scientists learned that it had a tiny but significant flaw. Because of faulty manufacturing procedures, the curve in its main mirror was off by just a fraction of a hair's width. Yet this flaw was enough to cause light to reflect away from the center of the mirror. As a result, the HST produced blurry images.

In spite of this handicap, the HST produced impressive beginning results. It was able to send back pictures of quasars such as the Einstein Cross, which is eight billion light-years away. (Because of the incredible vastness of space, astronomers and other scientists use the term light-year to refer to the distance lights travels in space in one year, approximately 6 trillion miles [9.6 trillion kilometers].) It also detected a white spot on Saturn, which turned out to be a storm system at least three times the size of Earth. Because computers could compensate for fuzzy images, the HST provided remarkable details about supernovas, the formation and merging of galaxies, the activity of black holes, and the composition of binary stars (a binary star is a pair of stars orbiting around one another, linked by gravity).

Fortunately, Hubble had been designed for regular maintenance by space shuttle crews. In early December 1993, astronauts aboard the space shuttle *Endeavour* completed repairs to the HST. They attached an apparatus called the Corrective

Optics Space Telescope Axial Replacement (COSTAR), a group of three quarter-sized mirrors, on the primary mirror. Like a pair of eyeglasses, the mirrors helped bring the light that the HST captured into proper focus. The astronauts also put in a new main camera and solar panels, and made other repairs. The total operation cost more than six hundred million dollars. One month after the repair mission, astronomers reported that virtually full vision had been restored to the HST.

In January 1994 HST embarked on an ambitious mission to search for black holes. By May of that year, it had uncovered evidence of a massive black hole—the size of three billion Suns—that is swallowing up matter in a galaxy near our own Milky Way. Two months later, the HST took hundreds of pictures as twenty large chunks of the comet Shoemaker-Levy 9 slammed into Jupiter, creating a 1,200-mile-wide (1,930-kilometer-wide) fireball that rose 600 miles (965 kilometers) above the planet's surface. It scarred the planet with a black dot about half the size of Earth. Taken after the impact points had rotated into view as seen from Earth, the images have helped astronomers learn more about the composition of comets and Jupiter and the dynamics of celestial crashes.

For ten consecutive days in December 1995, the HST pointed two of its cameras at a region of space covering one thirty-millionth of the sky. It was equivalent in apparent size to a shirt button held 75 feet (23 meters) away. The image compiled by the cameras over that period showed at least fifteen hundred faint galaxies. Astronomers believe that if this region of space, now known as the Hubble Deep Field, is typical of the rest of space, then hundreds of billions of galaxies, each containing billions of stars, exist within the known universe.

With the HST performing so well, NASA then embarked on a planned servicing mission in February 1997 to fine-tune the space-based observatory's instruments and to replace some of them with newer equipment. That month, a crew aboard the space shuttle *Discovery* made five spacewalks to service the HST in a three-hundred-million-dollar overhaul. All went well, ensuring the continued operation of the telescope.

Astronauts on two more servicing missions to the HST, conducted in December 1999 and March 2002, replaced faulty gyroscopes and fine guidance sensors, installed a new computer,

The purpose of the 17-ton Compton Gamma Ray Observatory, launched in 1991, was to study the universe at the wavelengths of gamma rays, the most energetic form of light. *(AP/Wide World Photos)*

replaced its solar panels, and upgraded several other instruments. On every service mission, the HST is boosted back into a higher orbit because atmospheric drag causes it to fall slowly out of orbit.

The HST was to have been serviced a final time by a planned space shuttle mission in February 2005. However, after the space shuttle *Columbia* disaster on February 1, 2003, all future shuttle service missions to the telescope were cancelled. Many astronomers were upset by the decision. They believed that the new manned space agenda proposed by U.S. president George W. Bush (1946–) in January 2004, which would send astronauts back to the Moon and to Mars, sealed the fate of the HST. Pressured by astronomers to keep the HST in orbit, NASA decided in June 2004 to explore a possible robotic repair mission to service the large space observatory by the end of 2007. Without repair work, the gyroscopes on the HST will begin failing by 2007. Its batteries are expected to die between 2008 and 2010. Sometime after that, the HST will reenter and burn up in Earth's atmosphere.

Compton: Exploring gamma rays

Cosmic gamma rays were first discovered in 1967 by small satellites called Velas. These military satellites had been put into orbit to monitor nuclear weapon explosions on Earth, but they found gamma ray bursts from outside the solar system, as well.

Several other small satellites launched by NASA in the early 1970s gave pictures of the whole gamma-ray sky. These pictures revealed hundreds of previously unknown stars and several possible black holes. Thousands more stars were discovered in 1977 and 1979 by three large satellites called High Energy Astrophysical Observatories (HEAO). They found that the entire Milky Way galaxy shines with gamma rays.

Then, on April 5, 1991, NASA launched the second of its Great Observatories, the Compton Gamma Ray Observatory (CGRO), into space aboard the space shuttle *Atlantis*. The telescope was named after U.S. physicist Arthur Holly Compton (1892–1962), who won the Nobel Prize in 1927 for his experimental efforts confirming that light had characteristics of both waves and particles.

The purpose of the 17-ton (15.4-metric ton) observatory, also known simply as Compton, was to study the universe at the wavelengths of gamma rays, the most energetic form of light. The CGRO carried four instruments to perform the necessary observations. Through data collected by the CGRO, astronomers have discovered that the center of the Milky Way glows in gamma rays created by the annihilation of matter and antimatter. (Antimatter is matter that is exactly the same as normal matter, but with the opposite spin and electrical charge. When matter and antimatter come into contact, both are annihilated with a tremendous release of energy.)

The Compton Effect

U.S. physicist Arthur Holly Compton (1892–1962), after whom the Compton Gamma Ray Observatory was named, shared the 1927 Nobel Prize in physics for his discovery of what came to be known as the Compton Effect.

In his research during the early 1920s, Compton noticed that when an X ray or gamma ray strikes an electron, it bounces off at an angle to its original trajectory and loses energy in the process. (An electron is a subatomic—smaller than an atom—particle that carries a single unit of negative electricity.) This loss of energy is demonstrated by the fact that the X ray or gamma ray consequently has a longer wavelength, a characteristic of its drop in speed. As the gamma ray data was less conclusive than the data on X rays, Compton limited his claims about this effect to X rays when he published the results of his research in 1923. Further research, however, demonstrated that the Compton Effect applied equally to gamma rays.

Compton's discovery was a major scientific breakthrough in determining that X

Arthur Holly Compton. *(AP/Wide World Photos)*

rays and gamma rays (and other forms of light) cannot be explained purely as a wave phenomenon. They must contain particles (or behave as if they do) in order to explain the Compton Effect. Physicists now speak of them as "wavy particles" because the subatomic particles do have wavelike characteristics, such as frequency and wavelength.

The CGRO also provided scientists with new information about supernovas, young star clusters, pulsars, black holes, quasars, solar flares, and gamma-ray bursts. Gamma-ray bursts are intense flashes of gamma rays that occur uniformly across the sky and thus likely originate from far outside the Milky Way. The energy of just one of these bursts has been calculated to be more than one thousand times the energy that the Sun will generate in its entire ten-billion-year lifetime. A major discovery of the CGRO was the class of objects called gamma-ray blazers: quasars that emit most of their energy as gamma rays and vary in brightness over a period of days.

Intended to operate for five years, the CGRO continued to work for several years beyond that period. In December 1999 NASA decided to end the CGRO's mission after one of its three gyroscopes, used to orient the observatory, failed. The CGRO, which cost more than six hundred million dollars, could have been kept aloft for eleven more years. Because the observatory was so heavy (it weighed even more than the HST), NASA was concerned that if the other gyroscopes failed, the CGRO could reenter the planet's atmosphere and crash, causing damage and injury. To prevent this, NASA deliberately brought the CGRO back into Earth's atmosphere on June 4, 2000 (after it had completed 51,658 orbits around the planet), where it broke apart. The charred remains of the observatory, roughly 6 tons (5.4 metric tons) of superheated metal, splashed into the Pacific Ocean about 2,500 miles (4,020 kilometers) southeast of Hawaii.

Chandra: Studying the universe at X-ray wavelengths

Since Earth's atmosphere absorbs the vast majority of X rays, they are not detectable from ground-based observatories, requiring space-based telescopes to make these observations. In 1970 NASA had launched *Uhuru* (Swahili for *freedom*), the first satellite designed specifically to study cosmic X-ray sources. By the time its mission ended in March 1973, *Uhuru* had produced a comprehensive map of the X-ray sky. The High Energy Astrophysical Observatories (HEAO), which NASA launched in 1977 to study gamma rays, also studied X rays. During its one-and-a-half years of operation, HEAO-1 provided constant monitoring of X-ray sources, such as individual stars, entire galaxies, and pulsars. The second HEAO, renamed the Einstein Observatory after it was launched, operated from November 1978 to April 1981. It contained a high-resolution X-ray telescope that discovered that X rays come from nearly every star.

In July 1999 NASA launched the Chandra X-ray Observatory (CXO), the third in its Great Observatories program. The observatory, originally called the Advanced X-ray Astrophysics

The Chandrasekhar Limit

Subrahmanyan Chandrasekhar (1910–1995) was an Indian-born U.S. astrophysicist and applied mathematician who proposed radical new theories of stellar evolution (referring to the changes that stars undergo during their "lifetimes"). His most celebrated work concerned the determination of the maximum mass of white dwarfs, which are the dying fragments of medium-sized stars. The Chandra X-ray Observatory was named in his honor.

In 1935 Chandrasekhar proposed the notion that as stars evolve, they emit energy generated primarily by their conversion of hydrogen into helium. As they reach the end of their lives, stars have progressively less hydrogen left to convert and thus emit less energy in the form of radiation. They eventually reach a stage when they are no longer able to generate the pressure needed to sustain their size against their own gravitational pull. At this point, stars begin to contract, or shrink.

As their density increases during the contraction process, they begin to collapse into themselves. Their electrons (subatomic particles that carry a single unit of negative electricity) become so tightly packed that the stars turn into tiny objects of enormous density: white dwarfs. According to Chandrasekhar, the greater the mass of a white dwarf, the smaller its radius. He went on to assert that not all stars end their lives as stable white dwarfs. If the mass of evolving stars increases beyond a certain limit, they cannot become stable white dwarfs. This

Subrahmanyan Chandrasekhar. *(AP/Wide World Photos)*

limit, calculated as 1.44 times the mass of the Sun, is now known as the Chandrasekhar limit. A star with a mass above the limit has either to lose mass to form a white dwarf or to take an alternative evolutionary path and become a supernova, which releases its excess energy in the form of a massive explosion.

Though at first ridiculed by other astronomers, Chandrasekhar's theory was later shown to be correct. Throughout his long career, he was recognized for his achievements with numerous awards and honors in the United States, Europe, and India. In 1983 he was awarded the Nobel Prize for physics for his research on the physical processes important to the structure and evolution of stars.

Facility, was renamed after the Nobel Prize-winning, Indian-born U.S. astrophysicist Subrahmanyan Chandrasekhar (1910–1995). It was launched into space by the space shuttle *Columbia* on mission STS-93. About one billion times more powerful than the first X-ray telescope, the CXO (sometimes known simply as Chandra) has a resolving power equal to the ability to read the letters of a stop sign at a distance of 12 miles (19 kilometers). This allows it to detect sources more than twenty times fainter than any previous X-ray telescope.

The CXO carries four instruments to study the universe at X-ray wavelengths, which are slightly less energetic than gamma rays. To carry out these observations, the CXO has an unusual orbit: Rather than moving in a circular orbit close to Earth, as was the case with the HST and the CGRO, it is in an elliptical or oval orbit that carries it between 6,200 and 86,800 miles (9,975 and 139,660 kilometers) away from the planet. This elliptical orbit allows the CXO to spend as much time as possible above the electrically charged particles in the Van Allen belts (two doughnut-shaped belts of high-energy charged particles trapped in Earth's magnetic field) that would interfere with its observations.

The purpose of the CXO is to obtain X-ray images of violent, high-temperature celestial events and objects to help astronomers better understand the structure and evolution of the universe. It will observe galaxies, black holes, quasars, and supernovae (among other objects) billions of light-years in the distance, giving astronomers a glimpse of regions of the universe as they existed eons ago. In early 2001 the CXO found the most distant X-ray cluster of galaxies astronomers have ever observed, located about ten billion light-years away from Earth. Less than a month later, it detected an X-ray quasar twelve billion light-years away.

During its relatively short time in orbit, the CXO has provided astronomers with a wealth of other data. Astronomers have used the observatory to learn more about the dark matter that may make up most of the mass of the universe, study black holes in great detail, witness the results of supernova explosions, and observe the birth of new stars. The CXO's mission was originally scheduled to last five years, but it will likely carry on as long as it continues to operate well.

Spitzer: The last of the Great Observatories

On August 25, 2003, NASA launched the Space Infrared Telescope Facility (SIRTF), the most sensitive instrument ever to look at the infrared spectrum in the universe. Astronomers had hoped it would peer through the veil of dust and gas that obscures most of the universe from view. Rather than go into Earth orbit, SIRTF was placed in an orbit around the Sun that has it trailing Earth by 5.4 million miles (8.7 million kilometers). This makes it easier for the observatory to perform observations without interference from Earth's own infrared light.

The first images taken by SIRTF, unveiled in December 2003, were designed to show off the abilities of the telescope. They showed a glowing stellar nursery where stars are born; a swirling, dusty galaxy; a disk of planet-forming debris; and organic material in the distant universe. Astronomers were ecstatic over the images. In keeping with NASA tradition, the observatory was renamed after its successful demonstration of operation. In honor of U.S. astrophysicist Lyman Spitzer, NASA officials changed the name of the space-based telescope to the Spitzer Space Telescope (SST).

The SST, which carries three instruments to detect infrared emissions, is scheduled for a five-year mission. During that time, astronomers plan to use the SST to study planets, comets, and asteroids in our solar system and to look for evidence of giant planets and brown dwarfs (small, cool, dark balls of matter that never complete the process of becoming a star) around other stars.

Other space observatories

Europe has been active in space exploration for decades. While individual European countries maintain national space programs, most space initiatives are a combined effort managed through the fifteen-nation European Space Agency (ESA), which was founded in 1962 as the European Space Research Organization. The countries that belong to ESA are Austria, Belgium, Denmark, Finland, France, Germany, Ireland, Italy, the Netherlands, Norway, Portugal, Spain, Sweden, Switzerland, and the United Kingdom.

ESA is a prime partner in the ongoing International Space Station program. It is also a partner with NASA in the Hub-

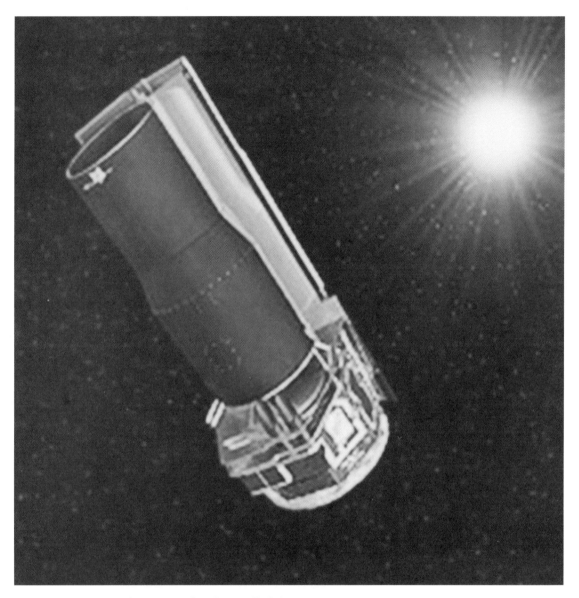

The Space Infrared Telescope Facility, later called the Spitzer Space Telescope, was launched in 2003 to look at the infrared spectrum in the universe. *(AP/Wide World Photos)*

ble Space Telescope, the Ulysses solar probe, several Earth-observation satellite systems, and several space-based observatories. Included in those is the Solar and Heliospheric Observatory (SOHO), which is studying the Sun.

Since its launch on December 2, 1995, SOHO has provided astronomers with unprecedented information about the Sun, from its interior to its hot and dynamic atmosphere to the solar wind and its interaction with the interstellar medium (the gas and dust that exists in the space between stars). The small observatory, which weighs only 4,080 pounds (1,850 kilograms), orbits the Sun in step with Earth. Its elliptical or oval orbit keeps it about 931,500 miles (1,500,000 kilometers) away from Earth (about four times the distance of the Moon), where the combined gravity of Earth and the Sun keep it locked. In this orbit, it has an uninterrupted view of the Sun, avoiding solar eclipses by Earth.

SOHO has given astronomers the first images ever of a star's turbulent outer shell and of the structure of sunspots below the surface (sunspots are cool areas of magnetic disturbance that form dark blemishes on the surface of the Sun). It has also provided the most detailed and precise measurements of the temperature structure and gas flows in the Sun's interior. In addition to observing the Sun, SOHO has discovered many comets. As of mid-2004, it had discovered eight hundred comets. Originally designed for a two-year mission, SOHO has been granted an extended life that will see it making observations through March 2007.

In addition to its joint projects, ESA has launched a number of its own observatories to study the universe. The Infrared Space Observatory (ISO), launched in 1995, gave astronomers unmatched views of the universe at infrared wavelengths. Among its most important discoveries is that disks of dust and gas, out of which planetary systems might form, surround a large fraction of young stars. ISO operated until 1998. In 1999 ESA launched XMM-Newton, an orbiting X-ray observatory similar to NASA's CXO. However, its mirror area and the energy range it can explore exceed that of CXO. At 3.8 tons (3.4 metric tons), it is the largest science satellite ever built in Europe. It has three advanced X-ray telescopes, each containing fifty-eight high-precision mirrors that offer the largest possible collecting area. In addition, it carries five X-ray imaging cameras and spectrographs. Its orbit around Earth is highly elliptical: It travels away to a distance of about 70,900 miles (114,000 kilometers), nearly one-third of the distance to the Moon. This allows it to undertake long, uninterrupted observations of faint X-ray sources.

NASA's Spitzer Space Telescope captured the dusty, star-filled M81, a spiral galaxy similar to our own, in this 2003 image. *(AP/Wide World Photos)*

The Canadian Space Agency has contributed a number of small observatories. In June 2003 the agency launched MOST (Microvariability and Oscillations of STars), Canada's first space telescope successfully put into space. It is also the smallest space telescope in the world: Suitcase-sized, it measures 25.6 inches by 25.6 inches by 11.8 inches (65 centimeters by 65 centimeters by 30 centimeters) and weighs 132 pounds (60 kilograms). The purpose of the microsatellite is to probe stars and extrasolar planets (planets in orbit around stars other than the Sun) by measuring tiny light variations undetectable from Earth. Two months later, the agency launched SciSat-1 (Science Satellite 1). The small, 330-pound (150-kilogram) observatory is an atmospheric research satellite designed to improve the understanding of the depletion of the ozone layer, with a special emphasis on the changes occurring over Canada and in the Arctic. Its mission is scheduled to run for two years.

Japan has also launched a few small space-based observatories. The Advanced Satellite for Cosmology and Astrophysics (ASCA) observatory was launched in 1993 and continues to operate in the early twenty-first century, studying the universe at X-ray wavelengths. Among its targets for study have been the cosmic microwave background radiation, galaxy clusters, and supernovae and their remnants. ASCA is also known by its national name, Asuka (Japanese for "flying bird"). The *Yohkoh* observatory, which operated between 1991 and 2001, carried four instruments that provided valuable data about the Sun's corona (the outermost and hottest layer of the Sun's atmosphere that extends out into space for millions of miles) and solar flares. In 1997 the Highly Advanced Laboratory for Communications and Astronomy (HALCA) satellite was placed in orbit. Through its 26-foot-diameter (8-meter-diameter) radio telescope, HALCA conducts joint observations with radio telescopes on Earth. Its elliptical orbit takes it as far away from Earth as 13,285 miles (21,375 kilometers). Because of this, it is also known by the national name Haruka (Japanese for "far away").

Future space-based observatories

The success of past and present space-based observatories has led NASA, ESA, and other space agencies to plan a new series of larger, more complex spacecraft that will be able to see deeper into the universe and in more detail than their predecessors. Leading these future observatories is the James Webb Space Telescope (JWST; previously called the Next Generation Space Telescope), named after NASA's second administrator, James E. Webb (1906–1992). Scheduled for launch sometime between 2009 and 2011, the JWST is intended (in part) to succeed the Hubble Space Telescope (HST). An infrared observatory, it will observe wavelengths between those at the red end of the visible spectrum and those at the middle of the infrared range. It will use a telescope up to 21.3 feet (6.5 meters) in diameter that will allow it to observe dimmer and more distant objects than the HST. Its primary mission will be to examine infrared remnants of the big bang, making observations of an earlier state of the universe than is possible at the present. The JWST will orbit at a distance of 931,500 miles (1,500,000 kilometers) away from Earth so that it can point permanently away from the infrared glow of Earth and the Sun.

To study gamma rays further, NASA (in association with government agencies in France, Italy, Japan, and Sweden) plans to launch the Gamma Ray Large Area Telescope (GLAST) in 2006.

ESA is also developing several space-based observatories that will study the universe at different electromagnetic wavelengths. Planck, scheduled for launch in 2007, will build upon the observations of the cosmic microwave background radiation made by COBE and WMAP. Also scheduled for launch that year will be Herschel (formerly called the Far Infrared and Submillimeter Telescope), which will observe the universe at far-infrared wavelengths. Measuring 23 feet (7 meters) high and 14 feet (4.3 meters) wide and weighing 3.58 tons (3.25 metric tons), it will be the largest space telescope of its kind when launched.

In the future, space-based observatories may consist of several spacecraft working together. Such orbiting arrays of telescopes could allow astronomers to obtain better images without the need to build extremely large and expensive single telescopes. One such observatory, called the Terrestrial Planet Finder (TPF), would combine images from several telescopes, each somewhat larger than the HST, to create a single image. A system of this type would make it possible for astronomers to observe planets the size of Earth orbiting other stars. TPF is tentatively scheduled for launch in 2014. NASA is also considering the launch of a similar observatory, called Constellation-X, which would use four X-ray telescopes orbiting in close proximity to each other to create the observing power of one giant telescope that would be one hundred times more powerful than any existing one.

For More Information

Books

Davies, John K. *Astronomy from Space: The Design and Operation of Orbiting Observatories.* Second ed. New York: Wiley, 1997.

Kerrod, Robin. *Hubble: The Mirror on the Universe*. Buffalo, NY: Firefly Books, 2003.

Naeye, Robert. *Signals from Space: The Chandra X-ray Observatory*. Austin, TX: Raintree Steck-Vaughn, 2001.

Schlegel, Eric M. *The Restless Universe: Understanding X-ray Astronomy in the Age of Chandra and Newton*. New York: Oxford University Press, 2002.

Web Sites

"CGRO Science Support Center." *NASA Goddard Space Flight Center.* http://cossc.gsfc.nasa.gov/ (accessed on August 19, 2004).

"Chandra X-ray Observatory." *Harvard-Smithsonian Center for Astrophysics.* http://chandra.harvard.edu/ (accessed on August 19, 2004).

"The Hubble Project." *NASA Goddard Space Flight Center.* http://hubble.nasa.gov/ (accessed on August 19, 2004).

HubbleSite. http://www.hubblesite.org/ (accessed on August 19, 2004).

"Orbital Telescopes." *Students for the Exploration and Development of Science.* http://www.seds.org/spider/oaos/oaos.html (accessed on August 19, 2004).

"Spitzer Space Telescope." *California Institute of Technology.* http://www.spitzer.caltech.edu/ (accessed on August 19, 2004).

14

Space Probes

On April 12, 1961, Soviet cosmonaut Yuri Gagarin (1934–1968) lifted off in *Vostok 1*, becoming the first human in space. His historic flight, in which he made one orbit around Earth, marked the beginning of manned spaceflight. In the more than four decades since then, programs launching humans into space have been carried out by the Soviet Union (later present-day Russia), the United States, and the People's Republic of China. More than 430 humans have flown into space. Most, though, have not flown beyond Earth orbit. Only the United States has carried out human spaceflight missions beyond Earth orbit, sending twenty-four astronauts to orbit and land on the Moon. The Moon revolves around Earth on an elliptical, or oval, orbit. The point in its orbit when it is farthest away from Earth, known as its apogee (pronounced AP-eh-gee), is about 252,780 miles (406,720 kilometers). As of 2004, this is the farthest humans have ventured out into space.

Manned spaceflight, with its awe-inspiring triumphs and heart-rending tragedies, dominated early space-travel news. The early exploration of space was a political race, pitting the Communist Soviet Union against the democratic, capitalist

Artist's rendition of the Near Earth Asteroid Rendezvous (NEAR) spacecraft, the only probe to have ever landed on an asteroid. *(AP/Wide World Photos)*

United States. The two countries were engaged in a Cold War, a prolonged conflict for world dominance that lasted from 1945 to 1991. A war that was fought both directly and indirectly, it influenced virtually every significant event or development in world affairs. Much of the Cold War was fought with propaganda, or information spread to further one's own cause. The victor of the space race, with its ultimate goal a landing on the Moon, would be able to brag about technological and political superiority.

Yet, while scientists and engineers from both nations struggled mightily to build rockets and capsules that would place either U.S. astronauts or Soviet cosmonauts first on the Moon, other scientists and engineers on both sides were pursuing the true dream of space exploration. In January 1959, more than two years before the first human escaped Earth's atmosphere, the Soviet Union's *Luna 1* flew within 3,725 miles (5,995 kilometers) of the Moon's surface, heralding the age of planetary exploration. This vehicle was a probe, an unmanned spacecraft that leaves Earth's orbit to explore the Moon, other celestial bodies, or outer space. Since its launch, more than one hundred other probes have been launched successfully on missions to obtain closer observations of the planets, their moons, the Sun, comets, asteroids, and the outer reaches of the solar system. As of the beginning of 2004, one probe, *Voyager 1,* had traveled more than 8.4 billion miles (13.5 billion kilometers) away from the Sun, slightly more than ninety times the distance between Earth and the Sun.

Probes have been sent beyond Earth orbit to pass near (called a flyby), orbit, or land on other celestial objects. No matter what their eventual destination, the primary objective of probes is to make scientific observations, such as taking pictures, analyzing atmospheric and soil conditions, measuring temperatures and magnetic fields (fields of force around the Sun and the planets generated by electrical currents), and collecting soil and rock samples. The information gathered by the probes is then either relayed or brought back to Earth.

To collect information, probes must carry with them some means of collecting and distinguishing this information. Sensors are one type of instrument used to perform this task. These instruments are programmed to detect alterations or variations in the space environment and send electrical, radio, or other types of signals or transmissions back to a main collection or recording device. Such a device may be aboard the spacecraft itself, on another spacecraft close by, or at a receiving station on Earth.

Time is a factor in the transmission of this information. To send a signal from Earth to the Moon and to receive one back takes about two seconds. For Mars, it takes between eighteen and forty-five minutes, depending on the relative orbiting positions of Mars and Earth. A signal from Pluto, the only planet in the solar system that has not had a spacecraft at least pass near it, would take five hours to reach Earth.

While some sensors gather information remotely about the conditions found in space or on a celestial body, other types of sensors may be used by a probe to make determinations about the position or location of the probe itself or its condition while in flight. Such sensors, onboard the probe and active during its flight, are essential elements in controlling the spacecraft or flying it to a specific destination in space.

Space probes must be able to last for years in space. While it takes the space shuttle only minutes to reach Earth orbit, it takes a probe a year to get to Mars. The U.S. probe *Galileo,* launched in October 1989, arrived at Jupiter through a complex route more than six years later. Once there, it began orbiting and sending back data about the planet and its moons for another eight years. Because there is no means of replacing or repairing parts while they are on a mission, probes have to be remarkably reliable. Interplanetary space (the area of space between the planets) has dangers that could destroy the sensors and other electrical workings of probes. Among these dangers are cosmic radiation (high-energy radiation coming from all directions in space), solar radiation, and the solar wind (electrically charged subatomic particles that flow out from the Sun). In addition, there is the potential for physical damage from dust or even larger chunks of floating material. Probes also have to be resilient to the extreme range of hot and cold temperatures to which they are exposed in space.

The Moon

In the first decade of the space race, the Soviet Union and the United States combined launched about fifty space probes

Words to Know

Apogee: The point in the orbit of an artificial satellite or the Moon that is farthest from Earth.

Artificial satellite: A man-made device that orbits Earth and other celestial bodies and that follows the same gravitational laws that govern the orbit of a natural satellite.

Cold War: A prolonged conflict for world dominance from 1945 to 1991 between the democratic, capitalist United States and the Communist Soviet Union. The weapons of conflict were commonly words of propaganda and threats.

Cosmic radiation: High-energy radiation coming from all directions in space.

Ecliptic: The imaginary plane of Earth's orbit around the Sun.

Escape velocity: The minimum speed that an object, such as a rocket, must have in order to escape completely from the gravitational influence of a planet or a star.

Flyby: A type of space mission in which the spacecraft passes close to its target but does not enter orbit around it or land on it.

Hard landing: The deliberate, destructive impact of a space vehicle on a predetermined celestial object.

Heliosphere: The vast region permeated by charged particles flowing out from the Sun that surrounds the Sun and extends throughout the solar system.

Interplanetary: Between or among planets.

Magnetic field: A field of force around the Sun and the planets generated by electrical charges.

Magnetosphere: The region of space around a celestial object that is dominated by the object's magnetic field.

Moonlet: A small natural or artificial satellite.

Probe: An unmanned spacecraft sent to explore the Moon, other celestial bodies, or outer space; some probes are programmed to return to Earth while others are not.

Radiation: The emission and movement of waves or atomic particles through space or other media.

Rover: A remote-controlled robotic vehicle.

Soft landing: The slow-speed landing of a space vehicle on a celestial object to avoid damage to or the destruction of the vehicle.

Solar wind: Electrically charged subatomic particles that flow out from the Sun.

to explore the Moon, the closest celestial target to Earth. The first probes were intended either to make a flyby or a hard landing (the deliberate, destructive impact of a space vehicle). Later probes achieved stable orbits around the Moon or made soft landings (the slow-speed landing of a space vehicle on a celestial object to avoid damage to or the destruction of the vehicle with sensors and other instruments intact). Each of these four objectives required increasingly greater rocket power and more precise maneuvering.

Between 1959 and 1976, the Soviet Union's Luna space probes thoroughly explored the Moon and space around it. This series of twenty-four probes accomplished a number of "firsts" in unmanned space exploration: They were the first human-made objects to reach escape velocity, which is the minimum speed that an object must have in order to escape completely from the gravitational influence of a planet or a star. They were also the first spacecraft to crash into the Moon, to photograph the Moon's farside (the side of the Moon that never faces Earth), to soft-land on the Moon, to return lunar soil to Earth, and to release a rover (a remote-controlled robotic vehicle) on the Moon's surface. Because the Soviets were very secretive about their early space programs, space experts have speculated the Luna program was intended to be a stepping-stone for manned lunar missions. This was a feat the Soviets were never able to achieve.

Fifteen of the Luna probes recorded successful missions. *Luna 1,* launched on January 2, 1959, was not among these. Although it was the first artificial satellite to travel beyond Earth's gravitational field (the force field created around massive bodies that causes attraction of other massive bodies), the primary objective of its mission was to hit the Moon. A failure in its control system caused it to miss its mark; instead, it only flew within about 3,725 miles (5,555 kilometers) of the Moon's surface. It was then eventually pulled into orbit around the Sun, becoming the first spacecraft to orbit the star at the center of the solar system. During its flyby of the Moon, *Luna 1* measured and reported that the Moon had no magnetic field.

Launched the following September, *Luna 2* became the first human-made object to land on the Moon when it made a hard landing east of the Sea of Serenity. (The "seas" on the

Moon are in fact large, dark lava plains formed by ancient volcanic eruptions caused by extremely large meteoroid impacts. Ancient astronomers, who mistook them for actual seas, dubbed them *maria,* Latin for "seas." They cover 16 percent of the lunar surface.) On impact, *Luna 2* scattered Soviet emblems and ribbons across the Moon's surface. *Luna 3* made a flyby of the farside of the Moon a month later. The pictures it took provided humans with the first view of this side of the Moon that is never visible from Earth.

What these early probes lacked was a propulsion system and a navigation system; they were simply thrown at the Moon. After *Luna 3,* the program was put on hold while Soviet engineers developed more sophisticated probes, ones capable of soft landings. As it turned out, this was not so easy. The next five Luna probes either self-destructed during launch, missed their target, or hard-landed on the Moon.

Success finally came with *Luna 9,* launched on January 31, 1966. The landing capsule was a 220-pound (100-kilogram) sphere that the probe dropped to the lunar surface to make history's first soft landing of a human-made object on the Moon. The capsule contained a television camera that sent back images to Earth. Although grainy, the pieced-together images provided the first detailed view of the lunar surface.

Of the next five Luna probes, only one, *Luna 13,* landed on the Moon. All the rest went into lunar orbit, studying conditions in space around the Moon, such as radiation and gravity, to determine how they might affect human travelers.

Luna 15, after having completed fifty-two orbits of the Moon, was to have soft-landed on the lunar surface on July 21, 1969, the day after *Apollo 11* astronauts Neil Armstrong (1930–) and Edwin "Buzz" Aldrin Jr. (1930–) had become the first humans on the Moon. However, the probe slammed into the surface at about 300 miles (480 kilometers) per hour. Had it made a successful landing, the probe was to have collected a sample of lunar soil and returned it to Earth. In September of the following year, *Luna 16* accomplished that task, returning 3.5 ounces (100 grams) of lunar soil and rock.

Luna 17 and *21,* launched in 1970 and 1973, respectively, each placed a rover on the Moon. The remote-controlled vehicles, called *Lunakhod 1* and *2,* were bathtub-shaped, measuring 8 feet (2.4 meters) long and 5.25 feet (1.6 meters) wide.

The launch of *Orbiter 1,* 1966. The probes in the Lunar Orbiter program were designed to photograph potential landing sites for the manned Apollo missions. *(National Aeronautics and Space Administration)*

Each had eight wheels and a lid made of solar cells. The first rover operated for about one year and the second for about two-and-one-half months. They cruised over the rocky terrain, taking photographs and measuring the chemical composition of the soil.

The last probe in the series, *Luna 24,* landed on the Moon on August 18, 1976. The third mission to return samples of the lunar surface, it brought back almost 6 ounces (175 grams) of Moon rock. This was the last lunar mission the Soviet Union (or its successor, present-day Russia) launched. From 1964 to 1970, while the Luna program was underway, the Soviet Union had launched another series of probes under the program named Zond. The first three Zond missions conducted flybys of Venus, Mars, and the Moon, respectively. *Zond 3* sent back pictures of the Moon's farside that were superior to those of *Luna 3. Zond 4* to *8,* large probes that weighed about 10,000 pounds (4,540 kilograms) each, were tests for a Soviet manned lunar mission. Each of these latter probes looped around the Moon without going into lunar orbit.

The first U.S. lunar probes, the Ranger series, were designed to obtain close-up images of the Moon's surface. Each probe was outfitted with six cameras. They were programmed to capture images of the lunar surface and send them back to Earth up to the moment the probe hard-landed on the Moon. Unfortunately, the early Ranger probes were not as successful as the early Luna probes. Equipment failures dogged the first six missions. After the launch failures of *Ranger 1* and *2* in 1961, *Ranger 3* missed the Moon by approximately 22,875 miles (36,800 kilometers) and went into orbit around the Sun (where it remains in the present-day). The fourth through sixth probes in the series, launched between 1962 and 1964, either crashed into the Moon or missed it altogether. In all cases, they failed to return any information to Earth. The last three in the program—*Rangers 7, 8,* and *9*—more than made up for the shortcomings of the first six. These missions, which took place in 1964 and 1965, transmitted a total of more than seventeen thousand detailed pictures. They greatly advanced scientific knowledge of the lunar surface.

Along with the Ranger program, two other U.S. probe programs helped lead the way to a manned lunar-landing mission: Lunar Orbiter and Surveyor. The probes in the Lunar Orbiter program were designed to photograph potential landing sites for the manned Apollo missions. Altogether, five Moon-orbiting Lunar Orbiter probes were launched in 1966 and 1967. The program's objective was met by *Lunar Orbiter 3* in February 1967. The remaining two flights were able to carry out further photography of the lunar surface for purely scientific purposes. In the end, the probes photographed 99 percent of the Moon, both its nearside and farside.

Between 1966 and 1968, the National Aeronautics and Space Administration (NASA) launched seven Surveyor probes to soft-land on the Moon. Once on the surface, the tripod-shaped Surveyors evaluated the lunar soil and environment. One of the first objectives of the program was to disprove the belief held by some scientists that the lunar surface was covered with a thick layer of dust. When *Surveyor 1* made a successful soft landing in the Ocean of Storms in June 1966, it did so in only 1.2 inches (3 centimeters) of dust. While *Surveyor 2* crashed and *Surveyor 4* lost contact with NASA's control center, *Surveyor 3, 5, 6,* and *7* landed at different sites and carried out experiments on the surface, including analyzing the chemical composition of the lunar soil. Overall, the Surveyor probes acquired almost 90,000 images from five landing sites.

The success of the Ranger, Lunar Orbiter, and Surveyor programs gave NASA officials the confidence to push forward with manned lunar landing missions. Between 1969 and 1972, six Apollo spacecraft carrying a total of eighteen astronauts landed on the Moon.

NASA did not send another vehicle to the Moon for twenty-two years. In 1994 the space agency sent the probe *Clementine* to orbit the Moon. Its primary objective was to test sensors and spacecraft components under extended exposure to the space environment and to make scientific observations of the Moon and a near-Earth asteroid. Measurements made by *Clementine* suggested that water ice existed at the lunar poles. Four years later, NASA launched the *Lunar Prospector.* Its orbit carried it not around the Moon's equator but around its poles. The probe was designed to investigate the Moon, providing scientists with a map of the surface composition and possible polar ice deposits and measurements of magnetic and gravity fields, among other findings. It was hoped the mission would improve understanding of the origin, evolution, current state, and resources of the Moon. In March 1998 NASA officials announced that the *Lunar Prospector* had discovered ice at both the north and south lunar poles, confirming *Clementine*'s findings. Mission scientists estimated that the total mass of ice the probe detected was about 6.6 billion

A Lunar Burial

U.S. planetary geologist Eugene M. Shoemaker (1928–1997) contributed greatly to space science exploration, particularly of the Moon. His research at Meteor Crater in Arizona in the late 1950s led to his appreciation of the role of asteroid and comet impacts as a fundamental process in the evolution of planets.

Shoemaker had a deep desire to travel into space, but health problems prevented him from becoming the first geologist on the Moon. Instead, he helped select and train the Apollo astronauts in lunar geology and impact cratering. He also appeared on television, giving geologic commentary while Apollo astronauts conducted Moon walks.

In 1992 Shoemaker was awarded the National Medal of Science, the highest scientific honor bestowed by the president of the United States. The following year, he was part of a leading comet-hunting team that discovered comet Shoemaker-Levy 9 and charted the object's breakup. Pieces of the comet slammed into Jupiter in July 1994—an unprecedented event in the history of astronomical observations.

Eugene M. Shoemaker. *(© Roger Ressmeyer/Corbis)*

On July 18, 1997, while carrying out research on impact craters in the Australian outback, Shoemaker was killed in a car accident. To honor him, NASA placed a small vial containing his ashes aboard the *Lunar Prospector.* Those ashes were scattered on the surface of the Moon when the probe made a controlled crash there on July 31, 1999, after completing its mission.

tons (6 billion metric tons) scattered in craters at both poles. On July 31, 1999, the probe struck the Moon in a controlled crash to look for further evidence of ice, but none was found.

In September 2003 the European Space Agency (ESA) launched *SMART-1,* the first European spacecraft designed to visit the Moon. It was estimated that the solar-powered, slow-moving probe would take sixteen months to travel to the Moon. Once there, the 815-pound (370-kilogram) probe will

Mercury

The closest object to the Sun, Mercury is a small, bleak planet. In the solar system, only Pluto is smaller. The planet is named for the Roman messenger god with winged sandals. It was given its name because it orbits the Sun quickly, in just eighty-eight days. Scientists knew little about Mercury beyond its size, orbit, and distance from the Sun until the space probe *Mariner 10* photographed the planet in 1975. Mariner probes, an early series of NASA interplanetary probes, were the first to return significant data on the surface and atmospheric conditions of Venus, Mars, and Mercury.

Launched on November 3, 1973, *Mariner 10* first approached the planet Venus in February 1974, then used the planet's gravitational field to send it around like a slingshot in the direction of Mercury. The journey to Mercury took seven weeks. On its first flight past the planet, *Mariner 10* came within about 437 miles (704 kilometers) of Mercury's surface. Photos that were taken by the probe revealed an intensely cratered, Moon-like surface.

Mariner 10 then went into orbit around the Sun before conducting two more flybys of Mercury on September 21, 1974, and on March 16, 1975. During its second flyby, at an altitude of 29,200 miles (47,000 kilometers), *Mariner 10* photographed the sunlit side of Mercury and its south polar region. On its final flyby, at a much closer altitude of 230 miles (327 kilometers), the probe took three hundred photographs and measured the planet's magnetic field. All onboard systems were shut down on March 24, 1975, after the probe's supply of fuel ran out. It was left to float in space.

Despite its successful mission, *Mariner 10* photographed only about 45 percent of Mercury's surface and only in moderate detail. As a consequence, there are still many questions about the history and evolution of the planet. To study the chemical composition of its surface, its geologic history, the nature of its magnetic field, the size and state of its core, and other features of the planet, NASA launched the MESSENGER

Scientists knew little about Mercury beyond its size, orbit, and distance from the Sun until the space probe *Mariner 10* photographed the planet in 1975. *(National Aeronautics and Space Administration)*

(MErcury Surface, Space ENvironment, GEochemistry and Ranging) probe in early August 2004. On its 4.9-billion-mile (7.9-billion-kilometer) journey, which will include fifteen loops around the Sun, the solar-powered MESSENGER will fly past Earth once, Venus twice, and Mercury three times before it will ease into orbit around Mercury in March 2011. It will remain in orbit around the planet for one Earth-year.

Venus

Named after the Roman goddess of love and beauty, Venus is the closest planet to Earth. The two have long been considered sister planets. The reason for this comparison is that they are similar in size, mass, and age. While astronomers could not see beneath Venus's thick cloud cover until recently, they assumed the planet would have seas and plant life like that on Earth. It is now known, however, that is not the case.

The first successful Venus probe was *Mariner 2,* launched by NASA on August 27, 1962. Less than four months after liftoff, it passed Venus at a distance of 21,610 miles (34,770 kilometers). The data *Mariner 2* relayed to Earth confirmed that Venus has a backward spin, a very high surface temperature, a thick atmosphere composed mostly of carbon dioxide, and no magnetic field. Once it completed its mission, *Mariner 2* went into orbit around the Sun, where it remains to this day.

Although the United States was the first to explore Venus, the former Soviet Union conducted an intensive, two-decade-long effort to explore the atmosphere and surface of the planet. The name given to the sixteen probes sent to Venus between 1961 and 1983 was Venera, Russian for "Venus." While the Venera program was initially unsuccessful, over the years it went on to record an impressive list of "firsts" about Venus. Venera spacecraft were the first to probe Venus's atmosphere, land on its surface, analyze its soil, and map and return pictures of its surface.

Venera 1, launched on February 12, 1961, became the first spacecraft to fly past Venus, but all contact with the probe was lost just seven days after its launch. *Venera 2* suffered the same fate. The third probe in the series, *Venera 3,* attempted to land on the planet, but communication with the probe was lost as it descended through the planet's atmosphere.

A measure of success was finally achieved by *Venera 4* when it reached Venus in October 1967. As it descended toward the planet's surface, the probe transmitted ninety-four minutes of data on the temperature, pressure, and chemical composition of Venus's atmosphere. About 15 miles (24 kilometers) above the surface, the probe was crushed by the intense pressure of the atmosphere.

Following *Venera 4,* the landing probes in the series were built stronger and with small parachutes that would enable them to reach the surface more quickly. Despite these changes, *Venera 5* and *6* met with fates similar to that of *Venera 4.* It was not until *Venera 7,* launched in August 1970, that the first successful landing of a spacecraft on another planet took place. It sent back data for thirty-five minutes during its descent and for another twenty-three minutes after it had reached the surface, although the signals it sent back were very weak.

Two years later, *Venera 8* built on the success of its predecessor. As it floated down through Venus's atmosphere, it measured variations in wind speed. Then for fifty minutes after landing, it transmitted data on the amount of sunlight reaching the surface (similar in illumination to an overcast day on Earth), as well as basic information on soil composition.

Venera 9 and *10,* identical probes each consisting of an orbiter and a lander, went into orbit around Venus in October 1975. After spending a month photographing cloud layers in the upper atmosphere, the probes both released their landers. Equipped with cameras, the landers sent back the first black-and-white images of the rock-strewn Venusian surface. The landers survived about fifty minutes before being destroyed by the heat and pressure.

The next four Venera probes reached Venus between December 1978 and March 1982. Each dropped landers to the surface. Between them, they measured the chemical composition of the atmosphere and surface rocks, confirmed the presence of lightning, and took the first color photographs of the surface.

The final two probes in the series, *Venera 15* and *16,* began orbiting Venus in October 1983. Their mission was to construct detailed maps of the planet's surface using radar (they bounced radio waves off the surface and recorded the echoes that were returned). Over their eight months of operation in orbit, the two probes mapped a large part of Venus's northern hemisphere.

In 1978, the year that the Soviets had launched *Venera 11,* NASA had sent two of its own probes to Venus. Known as *Pioneer Venus 1* and *2,* the probes were part of the Pioneer program, a diverse series of NASA spacecraft designed for lunar and interplanetary exploration (the Pioneer lunar program was never successful and was eventually abandoned). Once in orbit around Venus, *Pioneer Venus 1* studied the planet's atmosphere and mapped about 90 percent of its surface. In October 1992, after it had run out of fuel, the probe descended toward the surface and burned up in the atmosphere. *Pioneer Venus 2* carried four smaller probes that it released once in orbit. Each of the four probes was targeted at a different part of the planet. After they were released, the small probes measured atmospheric temperature, pressure, density, and chemical composition at various altitudes. Only one of the four survived after impact, transmitting data from the surface for sixty-seven minutes.

After the end of their Venera program, the Soviets sent two more probes to Venus. Launched one week apart in December 1984, *Vega 1* and *2* were programmed to be the first Soviet spacecraft to visit more than one celestial body. Each craft carried a Venus lander and a Halley's Comet flyby probe. *Vega 1* reached Venus on June 9, 1985, and dropped its lander to the surface. After touching down safely, the lander operated for two hours, relaying pictures and information about the composition of the soil. *Vega 1* also released a helium-filled balloon that hovered for two days about 30 miles (48 kilometers) above the planet's surface. During that time, the balloon was blown by the Venusian winds to a point about 6,200 miles (9.976 kilometers) from its original position. Instruments hanging from the balloon measured atmospheric temperature and pressure, as well as wind speeds. The entire lander-and-balloon sequence was repeated a few days later by *Vega 2.*

The twin Vega probes then circled Venus and used its gravitational force to propel them (a technique called gravity assist) on an course that took them close to Halley's Comet in March 1986. The probes returned photographs and analyzed ejected gas and dust from the comet.

In May 1989 NASA sent the probe *Magellan* to map the surface of Venus. Named after Portuguese explorer Ferdinand Magellan (1480–1521), the probe was launched from the space shuttle *Atlantis,* making it the first planetary explorer to be launched from a shuttle. *Magellan* circled the Sun one-and-one-half times before reaching Venus fifteen months later on August 10, 1990.

Over the next four years, *Magellan* used sophisticated radar equipment to survey 99 percent of the planet's surface. In this way, it created the most highly detailed map of Venus to date and produced images of such high quality that for the first time scientists could study the planet's geologic history. *Magellan* also measured Venus's gravitational field. The probe eventually entered the planet's atmosphere and burned up on October 12, 1994.

In late 2005 ESA plans to launch *Venus Express,* the agency's first mission to Venus. Once in orbit, the probe will perform a global investigation of the Venusian atmosphere, the first probe designed to do so.

Mars

The exploration of Mars has been an important part of space exploration. Since the 1960s, dozens of spacecraft, including orbiters, landers, and rovers, have been launched toward the planet named for the Roman god of war. The exploration of Mars, though, has come at a considerable financial cost: roughly two-thirds of all spacecraft destined for the planet have failed in one manner or another before completing or even beginning their missions. Many of those that failed did so due to technical incompetence. Others failed for no clear scientific reason.

The Soviet Union was the first nation to send an unmanned mission to Mars. After a number of unsuccessful attempts, they launched the *Mars 1* probe in late 1962, but lost contact with it after a few months. In 1971 they succeeded in putting *Mars 2* and *3* in orbit around the planet. Both of these craft carried landers that descended to the planet's surface. But in each case, radio contact was lost after about twenty seconds. Two years later, the Soviets sent out four more Mars probes, only one of which successfully transmitted data about the planet. *Mars 5* was able to establish an orbit around Mars, but operated for only a few days, returning images of a small portion of the Martian southern hemisphere.

In 1988 the Soviets renewed their interest in Mars with the Phobos program. They launched two identical spacecraft, *Phobos 1* and *2,* to study the planet and its moons Phobos and Deimos. Contact with *Phobos 1* was lost while it was en route to Mars, and *Phobos 2* failed just before it was set to release two landers on the surface of Phobos.

In the meantime, NASA was busy with its own exploration of the red planet. In 1964 it launched *Mariner 3* to make a flyby of the planet. However, its solar panels did not unfold properly. Unable to collect the Sun's energy for power, the probe soon died when its batteries ran out. It is now in orbit around the Sun. Its sister probe, *Mariner 4,* fared much better. On July 14, 1965, it flew within 6,120 miles (9,847 kilometers) of the Martian surface. It sent back twenty-one pictures of the planet, giving the first glimpse of its cratered surface.

The launch of *Viking 1,* Cape Canaveral, Florida. On July 20, 1976, the lander of *Viking 1* made the first successful soft landing on Mars.
(National Aeronautics and Space Administration)

The 1969 flybys of *Mariner 6* and *7* produced more than two hundred new images of Mars, as well as more detailed measurements of the composition and structure of its atmosphere and surface. From the data the probes sent back, scientists were able to determine that the planet's south polar cap was composed mostly of carbon dioxide.

Two years later, *Mariner 9* became the first spacecraft to orbit Mars. During its one year in orbit, the probe transmitted footage of an intense Martian dust storm as well as images of 90 percent of the planet's surface. *Mariner 9* confirmed that water had once flowed on Mars, but found no signs of recent geologic activity. The spacecraft also took photos of the planet's two small moons.

In 1976 the U.S. probes *Viking 1* and *Viking 2* had more direct encounters with Mars. Each Viking consisted of an orbiter and a lander. Once in orbit, the probes immediately began transmitting photos of the surface back to Earth, which mission controllers studied for possible landing sites. On July 20, 1976, the lander of *Viking 1* made the first successful soft landing on Mars. The lander's two cameras began operating minutes later. They showed rust-colored rocks and boulders (due to the presence of iron oxide) with a reddish sky above. *Viking 2* went into orbit around Mars that August. Its lander was released on the opposite side of the planet from the *Viking 1* lander, making a successful landing on September 3, 1976.

Over the next few years, the landers collected and analyzed soil samples from various areas of the planet. By the summer of 1980, the two Viking landers had sent back numerous weather reports and pictures of almost the entire surface of the planet. They found that the Martian atmosphere is made principally of carbon dioxide and, thus, is not capable of supporting human life. In addition, the soil they analyzed showed no signs of past or present life on the planet.

The orbiters and landers of both Viking probes operated far longer than anticipated. The last data received from the *Viking 2* lander was on April 11, 1980. After it was accidentally sent a wrong command, the *Viking 1* lander ceased operating on November 11, 1982.

Twenty years after the launch of the Viking probes, NASA launched the *Mars Global Surveyor* and the *Mars Pathfinder* to revisit Mars. After a three-hundred-day cruise through space, the *Mars Global Surveyor* reached Mars in September 1997. Orbiting the planet at an average altitude of 235 miles (378 kilometers), it mapped the entire surface of the planet. By the time it had completed its mapping mission in early 2001, it had sent back tens of thousands of images of the planet. It remains in orbit, functioning as a communications satellite to relay data back to Earth from surface landers of present and future Mars missions. After *Mars Pathfinder* landed on the planet on July 2, 1997, it released the first Martian rover: a miniature 22-pound (10-kilogram) vehicle called *Sojourner* (after Sojourner Truth [c. 1797–1883], a former U.S. slave who

Failure: Japan's First Interplanetary Probe

On July 3, 1998, the Japan Aerospace Exploration Agency (JAXA) launched its first interplanetary explorer. The probe's name was *Nozomi,* Japanese for "Hope." It was designed to study the Martian upper atmosphere and its interaction with the solar wind, the stream of highly charged particles coming from the Sun. The dragonfly-shaped probe, which weighed 1,190 pounds (541 kilograms), was to have entered an orbit at an altitude of about 550 miles (885 kilometers) above the planet's surface.

But in December 2003, JAXA scientists decided to end its mission. Malfunctions during the probe's journey had altered its trajectory, putting it into a course that was too low. They feared *Nozomi* might crash into the Martian surface, contaminating it with Earth bacteria. (Since the probe had not been designed to land, it had not been properly sterilized.) Solar flares had badly

Nozomi, Japan's first interplanetary explorer. *(AP/Wide World Photos)*

damaged its electrical and communications systems, and it was nearly out of fuel.

On December 9, 2003, small thrusters on *Nozomi* fired to take it out of its approach to Mars. Its new course put it in a two-year orbit around the Sun.

became a known antislavery speaker). During its three months of operation, *Mars Pathfinder* sent back more than sixteen thousand images from the lander and more than five hundred from the rover. It also sent back data on the chemical analyses of rocks, winds, and other aspects of the Martian weather.

Not all probes sent to Mars were as productive. In 1999 NASA lost two probes, the *Mars Climate Orbiter* and the *Mars Polar Lander*. Both were part of the Mars Surveyor '98 program. As their names imply, the *Mars Climate Orbiter* was to have explored the Martian atmosphere, while the *Mars Polar Lander* was to have landed near the planet's south polar cap to search for evidence of ice. Neither was able to land successfully. NASA officials believe that a software glitch caused the landing rockets on the *Mars Polar Lander* to shut down prematurely, causing the probe to slam into the surface at a high speed. An error in converting English and metric units of measurement in the navigation system of the *Mars Climate Orbiter* caused it to enter the Martian atmosphere at too low of an altitude, where it was destroyed.

In 2001 NASA was back on track when the *Mars Odyssey* probe settled into orbit around the planet in October of that year. Its mission was to hunt for past or present evidence of water and volcanic activity on Mars. In doing so, it would help determine whether the environment on Mars was ever conducive to life. It began mapping the planet in February 2002. Among its early findings was the discovery of ice deposits just underneath the soil near the planet's north pole. It also picked up signs that ice might be found at the south pole. In addition to its mapping duties, the probe acted as a relay for communication between landers on the planet and mission controllers on Earth.

ESA's first mission to Mars, the *Mars Express,* lifted off in early June 2003. The probe was to become the first spacecraft to use radar to penetrate the surface of Mars and to map any possible layers of water or ice. It also carried a lander, *Beagle 2,* named for the ship that carried English naturalist Charles Darwin (1809–1882) on his scientific voyages in the 1830s. The 143-pound (65-kilogram) lander, built by English scientists, was designed to use a robotic arm to gather and sample rocks for evidence of organic matter and water. The $345-million mission was to have lasted for one Martian year, or 687 Earth days. However, after *Beagle 2* was to have landed on Mars on December 25, 2003, ESA mission controllers did not hear any signal from it. Repeated attempts to contact the little lander proved unsuccessful, and its fate remains unknown. ESA officials believe that its parachute and air bags designed to cushion its landing may have been deployed too late or not at all. Despite this loss, the *Mars Express* continued its scientific mission. In January 2004 it detected ice at Mars's south pole, confirming the findings of the *Mars Odyssey.*

NASA continued its highly successful Mars Exploration Program, which began with the Viking landers and continued

An image of Phobos, the larger of Mars's two moons, as taken from the *Mars Global Surveyor*. *(National Aeronautics and Space Administration)*

through *Mars Pathfinder,* by placing two powerful new rovers on the planet in January 2004. The two rovers, *Spirit* and *Opportunity,* together made up the Mars Exploration Rover Mission. Each lifted off separately from Cape Canaveral, Florida: *Spirit* on June 10, 2003, and *Opportunity* on July 7, 2003. The six-month voyage to the red planet was uneventful, and *Spirit* landed almost exactly on target on Mars on January 3, 2004, followed three weeks later by *Opportunity* on January 25, 2004. The rovers landed on opposite sides of the planet, about 6,000 miles (9,650 kilometers) away from each other.

Each golf cart-sized rover weighed nearly 400 pounds (180 kilograms) and traveled across the Martian landscape on six wheels that moved independently, ensuring that they stayed in contact with the ground when the rover encountered rough terrain. The rovers moved in ten-second bursts, traveling at a slow rate that averaged about 120 feet (36.5 meters) per hour.

The goal of the Mars Exploration Rover Mission was to determine the history of climate and water at two sites on Mars where scientists believe conditions may have been favorable to life. To do so, each rover analyzed rocks and soil with a set of five geology instruments and an abrasion tool that exposed fresh rock surfaces for study. Each rover also carried a panoramic camera that sent images back to Earth to help mission controllers select the most promising targets on the surface for study.

In early March 2004 NASA officials announced that the rovers had found what they were looking for: evidence that Mars had once been a wet planet. Scientists determined this by studying the physical appearance and chemical composition of rocks that the rovers had found. The 820-million-dollar mission was declared a success when both rovers finished their ninety-day primary mission in April 2004. NASA officials then decided to let the rovers continue their explorations. By June 2004 *Opportunity* had traveled about 1 mile (1.6 kilometer) from its landing site while *Spirit* had traveled 2 miles (3.2 kilometers).

Jupiter

The largest planet in the solar system, Jupiter is thirteen hundred times larger than Earth. The fifth planet from the Sun, it is named after the chief Roman god, the god of light, the sky, and weather. With its sixty-three known moons, Jupiter is considered a mini-solar system of its own. Only one probe has been launched to orbit this large planet: *Galileo*. It was named in honor of Italian mathematician and astronomer Galileo Galilei (pronounced ga-lih-LAY-oh ga-lih-LAY-ee; 1564–1642), who discovered Jupiter's four largest moons—Io, Europa, Ganymede, and Callisto—in 1609.

The first probes to conduct flybys of Jupiter were *Pioneer 10* in 1973 and *Pioneer 11* in 1974. Of the two, *Pioneer 11* made the closest pass to the planet, 26,725 miles (43,000 kilometers). The suite of instruments aboard the probes made important observations about Jupiter's atmosphere and the space environment around the planet. Two more probes, *Voyager 1* and *2*, made flybys in 1979. They had been sent to build on the information acquired by the Pioneer probes. Their startling discoveries included finding a ring system around the planet and active volcanoes on its moon Io.

The NASA spacecraft that carried out the first studies of Jupiter's atmosphere, moons, and magnetosphere (the region of space around a celestial object that is dominated by the object's magnetic field) from orbit around the planet was *Galileo.* The probe had been launched from the space shuttle *Atlantis* in 1989. The nearly 3,000-pound (1,360-kilogram) *Galileo* began its journey in a direction opposite that of its destination. It headed first to Venus and looped around it, using that planet's gravitational field to propel it toward Jupiter. In all, *Galileo* traveled 2.5 billion miles (4 billion kilometers) to reach its target, which is about five times the distance between Earth and Jupiter.

Galileo finally went into orbit around Jupiter on December 7, 1995. On arrival, it dropped a barbeque-grill-sized mini-probe to the planet's surface. The mini-probe entered Jupiter's atmosphere at a speed of 107,025 miles (172,200 kilometers) per hour. Within two minutes, it had slowed to 100 miles (161 kilometers) per hour. Soon after, it released a parachute and began floating toward the surface. As it did so, intense winds blew it 300 miles (482 kilometers) horizontally. The mini-probe spent fifty-eight minutes collecting data on the weather of the gaseous planet before its cameras stopped working at an altitude of about 93 miles (150 kilometers) below the top of Jupiter's atmosphere. It was then either incinerated in the extreme heat of the atmosphere, with temperatures reaching 3,400°F (1,870°C), or crushed by atmospheric pressure.

Galileo's prime mission was to conduct a two-year study of Jupiter. However, because it operated so well, NASA mission controllers extended its mission three more times. In the end, it circled Jupiter for eight years, completing thirty-five orbits. The discoveries *Galileo* made were spectacular: It made the first observation of ammonia clouds in another planet's atmosphere. It also observed numerous large thunderstorms on Jupiter many times larger than those on Earth, with lightning strikes up to one thousand times more powerful. It was the first spacecraft to dwell in a giant planet's magnetosphere long enough to investigate the dynamics of Jupiter's magnetic field. *Galileo* determined that Jupiter's ring system is formed by dust kicked up as interplanetary meteoroids smash into the

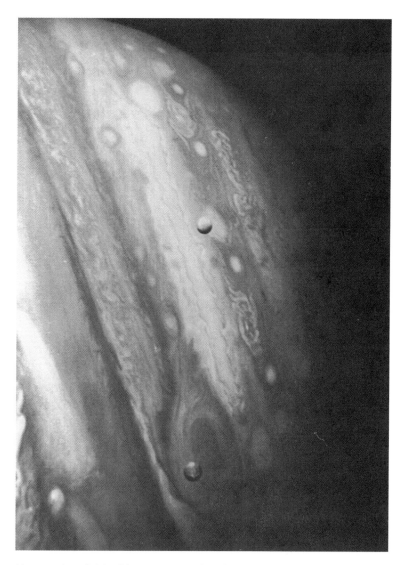

Voyager 1 and *2* had been sent to the planet Jupiter to build on the information acquired by the Pioneer probes. Their startling discoveries included finding a ring system around the planet and active volcanoes on its moon Io. *(National Aeronautics and Space Administration)*

planet's four small inner moons. Data from the probe also showed that Jupiter's outermost ring actually consists of two rings, one embedded within the other.

Many scientists believe that *Galileo*'s greatest discoveries were about the geologic diversity of Jupiter's four largest moons. *Galileo* found that Io's extensive volcanic activity is one hundred times greater than that found on Earth. Its measurements of Europa indicated that beneath the moon's cracked crust of ice could be a salty ocean up to 62 miles (100 kilometers) deep. If so, it would contain about twice as much water as all of Earth's oceans. Data gathered by the probe also showed that Ganymede and Callisto may have a saltwater layer. The biggest discovery surrounding Ganymede was the presence of a magnetic field—no other moon of any planet is known to have one.

Galileo's hugely successful fourteen-year mission came to an end on September 21, 2003, when mission controllers decided to send it into Jupiter's stormy atmosphere. With its propellant running low and its electrical systems fading, *Galileo* plunged into the planet's atmosphere at a speed of 108,000 miles (173,770 kilometers) per hour, quickly disintegrating and bringing to an end its 1.4-billion-dollar mission.

Saturn

The sixth planet from the Sun, Saturn is named for the Roman god of agriculture. The second largest planet in the solar system, it is also the least dense of all planets. It is almost 30 percent less dense than water; placed in a large-enough body of water, Saturn would float.

Saturn's most outstanding characteristic is its rings. The three other largest planets—Jupiter, Uranus, and Neptune—also have rings, but Saturn's are the most spectacular. For centuries, astronomers believed that the rings were moons. In 1658 Dutch astronomer Christiaan Huygens (1629–1695) first identified the structures around Saturn as a single ring. In later years, equipped with increasingly stronger telescopes, astronomers increased the number of rings they believed surrounded the planet.

In 1980 and 1981 the *Voyager 1* and *Voyager 2* probes sent back the first detailed photographs of Saturn and its spectacular rings. The probes revealed a system of more than one thousand ringlets encircling the planet at a distance of 50,000 miles (80,450 kilometers) from its surface. *Voyager 1*'s closest approach to Saturn came on November 12, 1980, when it flew within 77,000 miles (123,890 kilometers) of the planet's cloud

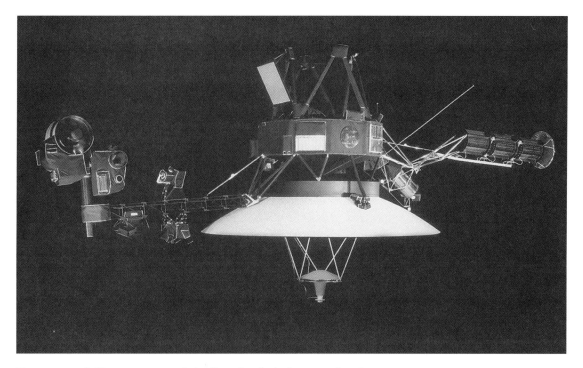

The spacecraft *Voyager* captured the first detailed photographs of Saturn and its spectacular rings. *(National Aeronautics and Space Administration)*

tops. It discovered fast-moving clouds in the planet's atmosphere and several previously unknown small moons. It also made a flyby of Titan, Saturn's largest moon. *Voyager 1* revealed that Titan may have seas of liquid methane bordered by organic tar-like matter. *Voyager 2* did not conduct as detailed an examination of Saturn as its sister probe, using its approach to the planet as a gravity-assist to send it on to Uranus and Neptune.

The main mission to study Saturn is a joint venture between NASA and ESA. On October 15, 1997, the two agencies launched the *Cassini-Huygens* spacecraft, which is composed of the NASA-built *Cassini* orbiter and the ESA-built *Huygens* probe. (The orbiter was named for the Italian astronomer Gian Domenico Cassini [1625–1712], who observed Saturn's rings and discovered four of its moons; the probe was named for Christiaan Huygens.) The spacecraft is one of the largest, heaviest, and most complex interplanetary spacecraft ever built. At launch, the total vehicle weighed about 12,350 pounds (5,600 kilograms). It stood more than 22.3 feet (6.8 meters) high and was more than 13.1 feet (4 meters) wide.

Costing about three billion dollars, the mission is the last of NASA's big-budget, big-mission planetary probes. Carrying twelve scientific instruments, the orbiter was planned to take a variety of measurements of Saturn's atmosphere, its moons, and the dust, rock, and ice that comprise its rings. After traveling some 2.2 billion miles (3.5 billion kilometers), the orbiter cruised into an orbit around the planet on June 30, 2004. A few weeks before then, it completed a flyby of Phoebe, the planet's largest outer moon. Coming within 1,285 miles (2,068 kilometers) of the dark moon, which measures just 137 miles (220 kilometers) across, *Cassini* took high-resolution photographs of the moon's deeply cratered surface.

The orbiter is scheduled to drop the *Huygens* probe onto the surface of Titan on December 25, 2004, for a detailed look at the moon's surface. The probe is expected to take twenty-one days to reach Titan. During its descent, *Huygens*'s camera will capture more than one thousand images, while the probe's other five instruments will sample Titan's atmosphere and determine its composition. If it survives the impact of its landing, *Huygens* will transmit data from its instruments back to *Cassini*. Over a four-year period, *Cassini* will orbit Saturn seventy-four times, make forty-four flybys of Titan, and make numerous flybys of the planet's other moons. It will send back as many as five hundred thousand color images taken with an onboard camera.

Uranus

Uranus was the first planet to be discovered that had not been known since ancient times. Although Uranus is just bright enough to be seen with the naked eye, and in fact had appeared in some early star charts as an unidentified star, English astronomer William Herschel (1738–1822) was the first to recognize it as a planet in 1781. The planet was named after Ouranos, the Greek god of the sky.

Most of what is known about Uranus was discovered during the 1986 *Voyager 2* flyby of the planet. *Voyager 2* had lifted off from Earth in August 1977; it first visited Jupiter in July 1979, then Saturn in August 1981.

At its closest approach, on January 24, 1986, *Voyager 2* came within 50,600 miles (81,415 kilometers) of the planet. Among its most important findings were ten previously undiscovered moons and two new rings. (The original nine rings of Uranus were discovered only nine years before the probe's visit. Since the probe's flyby, astronomers have found an additional twelve moons, bringing the total of known moons to twenty-seven.) *Voyager 2* determined that the five largest moons are made mostly of ice and rock. Some are heavily cratered, others have steep cliffs and canyons, and still others are much flatter.

Voyager 2 also made the first accurate determination of Uranus's rate of rotation (17.2 hours) and found a large and unusual magnetic field, one that is fifty times stronger than that of Earth. Finally, it discovered that despite greatly varying exposure to sunlight, the planet is about the same temperature all over: roughly –346°F (–210°C).

Neptune

Neptune is a large planet, seventeen times more massive than Earth and far more blue. Since it has a rich blue-green color, Neptune was named for the Roman god of the sea. This color is due to the presence of methane gas in its atmosphere (not water on its surface, like Earth).

Neptune is never visible to the naked eye. It was discovered in the 1840s only after astronomers deduced the presence of another planet from the shape of Uranus's orbit. To date, only one probe has visited Neptune: the workhorse *Voyager 2*. On August 25, 1989, the probe conducted a flyby of the gassy planet. It found that Neptune is encircled by at least four very faint rings, much less pronounced than the rings of Saturn, Jupiter, or Uranus. Although astronomers are not quite sure, they believe these rings are composed of particles, some of which measure more than 1 mile (1.6 kilometers) across and are considered moonlets (small natural or artificial satellites). These particles clump together in places, creating relatively bright arcs. This originally led astronomers to believe that only arcs—and not complete rings—were all that surrounded the planet.

Voyager 2 also discovered six of Neptune's eight known moons. When it flew by the planet, the probe detected nu-

Space Junk

The space around Earth is filled with thousands of pieces of junk ranging from nuts and bolts to entire satellites. The oldest debris still in orbit is the second U.S. satellite ever launched, *Vanguard 1,* which lifted off on March 17, 1958. Although it operated for only six years, it remains in its position floating around the planet. In 1965, during the first U.S. spacewalk, *Gemini 4* astronaut Edward H. White II (1930–1967) lost a glove. For one month, the glove stayed in orbit, traveling around Earth at a speed of about 17,500 miles (28,000 kilometers) per hour. More than two hundred objects, most of them rubbish bags, were released by cosmonauts aboard the space station *Mir* during its first ten years of operation.

At the beginning of the twenty-first century, the U.S. Space Command estimated that there was approximately four million pounds of space junk in low-Earth orbit. The agency counted 8,927 human-made objects immediately around the planet and beyond in space. Of the total, 2,671 were satellites (some working, some not), 90 were space probes that have been launched out of Earth orbit, and 6,096 were mere chunks of debris zooming around the planet. The United States leads all nations in the total quantity of orbital junk, but some companies and other organizations have contributed significantly to the count.

merous cloud features. The biggest was the Great Dark Spot, a hurricane-like storm that was about half the size of Earth. The next feature discovered was a small white spot, which appeared to race rapidly around the planet when compared with the slow-moving Great Dark Spot. The mysterious white spot was named Scooter. In 1994, however, observations from the Hubble Space Telescope showed that the Great Dark Spot had disappeared. Astronomers theorize that the spot either simply dissipated or is being masked by other aspects of the atmosphere. A few months later, the Hubble Space Telescope discovered a new dark spot in Neptune's northern hemisphere. This discovery has led astronomers to conclude that the planet's atmosphere, in which blow some of the fiercest winds in the solar system, changes rapidly.

After its close encounter with Neptune, *Voyager 2* joined its sister probe, *Voyager 1,* exploring the outer reaches of the solar system, where the Sun's influence ends and the dark recesses of interstellar space begin.

Comets, asteroids, and the Sun

Some of the planetary probes that have been launched have explored other celestial objects in addition to their primary planetary targets. Others have been designed to focus solely on those objects whose journey around the Sun is wildly eccentric. Among these objects are comets (relatively small, icy objects that travel around the Sun in a highly elliptical, or oval, orbit) and asteroids (medium-sized rocky bodies that orbit the Sun).

Perhaps the best-known comet is Halley's Comet, named after English astronomer Edmond Halley (pronounced HAL-ee; 1656–1742) who first suggested in 1682 that the comet completes one orbit around the Sun approximately once every seventy-six years. In 1986, when it was scheduled to pass near the Sun and Earth, it attracted a great deal of attention among both scientists and the general public. The Soviet probes *Vega 1* and *2*, the Japanese probes *Sakigake* and *Suisei*, and the European probe *Giotto* were all sent to observe the comet. Of these probes, *Giotto* made the most significant findings.

Built by ESA, *Giotto* came within 370 miles (596 kilometers) of the comet's center, capturing fascinating images of the 9-mile-long, 5-mile-wide (15-kilometer-long, 8-kilometer-wide) potato-shaped core marked by hills and valleys. Two bright jets of dust and gas, each 9 miles (15 kilometers) long, shot out of the core. *Giotto's* instruments detected the presence of water, carbon, nitrogen, and sulfur molecules. It also found that the comet was losing about 30 tons (27 metric tons) of water and 5 tons (4.5 metric tons) of dust each hour. This means that although the comet will survive for hundreds more orbits, it will eventually disintegrate. Halley's Comet is due to pass by Earth in the year 2061.

After *Giotto*, other probes have been launched to explore comets entering Earth space. In October 1998 NASA launched the *Deep Space 1* probe to make flybys of the comet Borrelly and the asteroid Braille. It completed both tasks successfully, coming within 16 miles (26 kilometers) of the asteroid. In February 1999 NASA sent the probe *Stardust* to investigate the makeup of the comet Wild 2 (pronounced Vilt 2). The primary goal of the probe was to collect dust and other material from the comet and return it to Earth. It was also programmed to collect samples of interstellar dust. In January 2004 *Stardust* flew within 149 miles (240 kilometers) of the comet, catching samples of comet particles in its dust collector grid that opened like a clamshell. It also captured clear images of the strangely shaped comet. The pictures showed that Wild 2 has towering peaks and steep-walled craters that seem to defy gravity. More than twelve jets of material shoot out from inside the comet. Dust swirls around it in dense packets. The material *Stardust* collected is expected to be returned to Earth in a capsule in 2006.

The early race into space, with its awe-inspiring triumphs and heart-rending tragedies, received much media coverage all over the world.

(© Bettmann/Corbis)

On March 2, 2004, ESA launched the *Rosetta* probe. Scientists named it after the Rosetta stone tablet that helped archeologists decipher Egyptian hieroglyphics. They hope the probe will reveal clues about the birth of the Sun and the planets of the solar system. It will do this by studying the comet called 67P/Churymov-Gerasimenko, which had been discovered by two Soviet astronomers in 1969. (Among the solar system's most primitive objects, comets are believed to hold deep-frozen matter left over from the birth of the Sun and the planets.) *Rosetta* is expected to reach the comet in May 2014, then go into orbit around it. Six months later, it will release the lander named *Philae* that will try to touch down on the surface of the comet. All previous spacecraft have only made brief flybys of comets.

The only probe to have ever landed on an asteroid was the *NEAR Shoemaker*. (It was originally named the Near Earth Asteroid Rendezvous, but was renamed after launch to honor U.S. planetary geologist Eugene M. Shoemaker [1928–1997].) In April 2000, after having traveled some 2 billion miles (3.2 billion kilometers) since it left Earth on February 17, 1996, the probe began a circular orbit around the asteroid Eros. It was the first time a spacecraft had orbited an asteroid. During its one-year mission around Eros, the 1,100-pound (500-kilogram) spacecraft settled into an orbit that at one point was as close as 3 miles (5 kilometers) above the potato-shaped asteroid. Eros, named after the Greek god of physical love, is one of the larger asteroids in the solar system, measuring about 21 miles (34 kilometers) long and 8 miles (13 kilometers) thick. It is called a near-Earth asteroid because its orbit crosses that of Earth and poses a potential collision danger. Scientists estimate that it is 4.5 billion years old, almost unchanged since the beginning of the solar system.

Loaded with six instruments, *NEAR Shoemaker* took measurements to determine the mass, density, chemical composition, and other geological characteristics of the asteroid. It also beamed back to Earth some 160,000 images of Eros. On February 12, 2001, *NEAR Shoemaker* used the last of its fuel in an attempt to land on the surface of the asteroid. The craft had not been designed with landing gear, and mission scientists had given it a 1 percent chance of survival. Bumping into the asteroid at a mere 4 miles (6.4 kilometers) per hour, however, the hardy spacecraft survived, becoming the first spacecraft to land on an asteroid. On its way down to the surface, *NEAR Shoemaker* continued to transmit pictures back to Earth. Once on the surface, it collected invaluable data about the oddly shaped asteroid. Even though scientists will probably study the data for years, they did learn early on that the asteroid does not tumble wildly through space. Instead, it rotates around one axis, much like a planet.

An end to its successful five-year mission came on February 28, 2001, when scientists put *NEAR Shoemaker* into a planned hibernation. They did not believe that the spacecraft would survive the frigid darkness of winter on Eros, when temperatures would plummet to –238°F (–150°C).

In 1990 NASA and ESA joined forces to deploy a probe that would make the first-ever measurements of the activity at the Sun's north and south poles and in the unexplored region of space above and below the poles. That probe, *Ulysses,* was launched from the cargo bay of the space shuttle *Discovery* on October 6, 1990. In order to study the Sun's poles, the probe had to cross out of the ecliptic, the imaginary plane of Earth's orbit around the Sun. (The orbits of all the major planets except Pluto lie near this plane.) The reason for this is that the Sun's poles can only be studied from above or below the Sun, points outside of the two dimensions of the ecliptic.

Ulysses initially headed away from the Sun, toward Jupiter. It then looped around Jupiter in February 1992 and used the giant planet's gravitational field to propel itself southward, out of the ecliptic. In September 1994 *Ulysses* crossed beneath the Sun's south pole and began heading north. One year later, it passed over the Sun's north pole. *Ulysses* then headed back toward Jupiter on the long leg of its six-year, oval-shaped path. It passed over the solar poles once again in 2000 and 2001.

Ulysses has provided scientists with the very first all-around map of the heliosphere, the vast region filled with charged particles flowing out from the Sun that surrounds the Sun and extends throughout the solar system. New facts about the fast solar wind, those flowing charged particles, were among the probe's most fundamental discoveries. The typical solar wind emerging from the Sun's equatorial, or middle area, is variable but relatively slow, at 220 to 250 miles (350 to 400 kilometers) per second. The fast solar wind blows at a steady 465 miles (750 kilometers) per second. It comes from fairly small, cool regions of the solar atmosphere that are close to the poles. Yet *Ulysses* found that the fast solar wind fans out to fill two-thirds of the volume of the heliosphere. The boundary between the two wind streams is also unexpectedly distinct.

Because it has proved to be so useful, scientists plan to have *Ulysses* operate until early 2008.

For More Information

Books

Benson, Michael. *Beyond: Visions of the Interplanetary Probes.* New York: Abrams, 2003.

Bredeson, Carmen. *NASA Planetary Spacecraft: Galileo, Magellan, Pathfinder, and Voyager.* Berkeley Heights, NJ: Enslow, 2000.

Hamilton, John. *The Viking Missions to Mars.* Edina, MN: Abdo and Daughters Publishing, 1998.

Kluger, Jeffrey. *Moon Hunters: NASA's Remarkable Expeditions to the Ends of the Solar System.* New York: Simon and Schuster, 2001.

Kraemer, Robert S. *Beyond the Moon: A Golden Age of Planetary Exploration, 1971–1978.* Washington, DC: Smithsonian Institution Press, 2000.

Sherman, Josepha. *Deep Space Observation Satellites.* New York: Rosen Publishing Group, 2003.

Web Sites

"Cassini-Huygens: Mission to Saturn and Titan." *Jet Propulsion Laboratory, California Institute of Technology.* http://saturn.jpl.nasa.gov/index.cfm (accessed on August 19, 2004).

"ESA: Space Science." *European Space Agency.* http://www.esa.int/export/esaSC/index.html (accessed on August 19, 2004).

"Galileo: Journey to Jupiter." *Jet Propulsion Laboratory, California Institute of Technology.* http://www2.jpl.nasa.gov/galileo/ (accessed on August 19, 2004).

"Mars Exploration Rover Mission." *Jet Propulsion Laboratory, California Institute of Technology.* http://marsrovers.jpl.nasa.gov/home/index.html (accessed on August 19, 2004).

"NASA: Robotic Explorers." *National Aeronautics and Space Administration.* http://www.nasa.gov/vision/universe/roboticexplorers/index.html (accessed on August 19, 2004).

"Voyager: The Interstellar Mission." *Jet Propulsion Laboratory, California Institute of Technology.* http://voyager.jpl.nasa.gov/ (accessed on August 19, 2004).

Where to Learn More

Books

Aaseng, Nathan. *The Space Race*. San Diego, CA: Lucent, 2001.

Andronik, Catherine M. *Copernicus: Founder of Modern Astronomy*. Berkeley Heights, NJ: Enslow, 2002.

Asimov, Isaac. *Astronomy in Ancient Times*. Revised ed. Milwaukee: Gareth Stevens, 1997.

Aveni, Anthony. *Stairways to the Stars: Skywatching in Three Great Ancient Cultures*. New York: John Wiley and Sons, 1997.

Baker, David. *Spaceflight and Rocketry: A Chronology*. New York: Facts on File, 1996.

Benson, Michael. *Beyond: Visions of the Interplanetary Probes*. New York: Abrams, 2003.

Bille, Matt, and Erika Lishock. *The First Space Race: Launching the World's First Satellites*. College Station, TX: Texas A&M University Press, 2004.

Bilstein, Roger E. *Orders of Magnitude: A History of the NACA and NASA, 1915–1990*. Washington, DC: National Aeronautics and Space Administration, 1989.

Boerst, William J. *Galileo Galilei and the Science of Motion*. Greensboro, NC: Morgan Reynolds, 2003.

Bredeson, Carmen. *NASA Planetary Spacecraft: Galileo, Magellan, Pathfinder, and Voyager.* Berkeley Heights, NJ: Enslow, 2000.

Caprara, Giovanni. *Living in Space: From Science Fiction to the International Space Station.* Buffalo, NY: Firefly Books, 2000.

Catchpole, John. *Project Mercury: NASA's First Manned Space Programme.* New York: Springer Verlag, 2001.

Chaikin, Andrew L. *A Man on the Moon: The Voyages of the Apollo Astronauts.* New York: Penguin, 1998.

Christianson, Gale E. *Edwin Hubble: Mariner of the Nebulae.* Chicago, IL: University of Chicago Press, 1996.

Clary, David A. *Rocket Man: Robert H. Goddard and the Birth of the Space Age.* New York: Hyperion Press, 2003.

Cole, Michael D. *The Columbia Space Shuttle Disaster: From First Liftoff to Tragic Final Flight.* Revised ed. Berkeley Heights, NJ: Enslow, 2003.

Collins, Michael. *Carrying the Fire: An Astronaut's Journeys.* New York: Cooper Square Press, 2001.

Davies, John K. *Astronomy from Space: The Design and Operation of Orbiting Observatories.* Second ed. New York: Wiley, 1997.

Dickinson, Terence. *Exploring the Night Sky: The Equinox Astronomy Guide for Beginners.* Buffalo, NY: Firefly Books, 1987.

Dickson, Paul. *Sputnik: The Shock of the Century.* New York: Walker, 2001.

Ezell, Edward Clinton, and Linda Neuman Ezell. *The Partnership: A History of the Apollo-Soyuz Test Project.* Washington, DC: National Aeronautics and Space Administration, 1978.

Florence, Ronald. *The Perfect Machine: Building the Palomar Telescope.* New York: HarperCollins, 1994.

Fox, Mary Virginia. *Rockets.* Tarrytown, NY: Benchmark Books, 1996.

Gleick, James. *Isaac Newton.* New York: Pantheon Books, 2003.

Hall, Rex, and David J. Shayler. *The Rocket Men: Vostok and Voskhod, the First Soviet Manned Spaceflights.* New York: Springer Verlag, 2001.

Hall, Rex D., and David J. Shayler. *Soyuz: A Universal Spacecraft.* New York: Springer Verlag, 2003.

Hamilton, John. *The Viking Missions to Mars.* Edina, MN: Abdo and Daughters Publishing, 1998.

Harland, David M. *The MIR Space Station: A Precursor to Space Colonization.* New York: Wiley, 1997.

Harland, David M., and John E. Catchpole. *Creating the International Space Station.* New York: Springer Verlag, 2002.

Holden, Henry M. *The Tragedy of the Space Shuttle Challenger.* Berkeley Heights, NJ: MyReportLinks.com, 2004.

Jenkins, Dennis R. *Space Shuttle: The History of the National Space Transportation System.* Third ed. Cape Canaveral, FL: D. R. Jenkins, 2001.

Kerrod, Robin. *The Book of Constellations: Discover the Secrets in the Stars.* Hauppauge, NY: Barron's, 2002.

Kerrod, Robin. *Hubble: The Mirror on the Universe.* Buffalo, NY: Firefly Books, 2003.

Kluger, Jeffrey. *Moon Hunters: NASA's Remarkable Expeditions to the Ends of the Solar System.* New York: Simon and Schuster, 2001.

Kraemer, Robert S. *Beyond the Moon: A Golden Age of Planetary Exploration, 1971–1978.* Washington, DC: Smithsonian Institution Press, 2000.

Krupp, E. C. *Beyond the Blue Horizon: Myths and Legends of the Sun, Moon, Stars, and Planets.* New York: Oxford University Press, 1992.

Launius, Roger D. *Space Stations: Base Camps to the Stars.* Washington, DC: Smithsonian Institution Press, 2003.

Maurer, Richard. *Rocket! How a Toy Launched the Space Age.* New York: Knopf, 1995.

Miller, Ron. *The History of Rockets.* New York: Franklin Watts, 1999.

Murray, Charles. *Apollo: The Race to the Moon.* New York: Simon and Schuster, 1989.

Naeye, Robert. *Signals from Space: The Chandra X-ray Observatory.* Austin, TX: Raintree Steck-Vaughn, 2001.

Orr, Tamra B. *The Telescope.* New York: Franklin Watts, 2004.

Panek, Richard. *Seeing and Believing: How the Telescope Opened Our Eyes and Minds to the Heavens.* New York: Penguin, 1999.

Parker, Barry R. *Stairway to the Stars: The Story of the World's Largest Observatory.* New York: Perseus Publishing, 2001.

Reichhardt, Tony, ed. *Space Shuttle: The First 20 Years—The Astronauts' Experiences in Their Own Words.* New York: DK Publishing, 2002.

Reynolds, David. *Apollo: The Epic Journey to the Moon.* New York: Harcourt, 2002.

Ride, Sally. *To Space and Back.* New York: HarperCollins, 1986.

Shayler, David J. *Gemini: Steps to the Moon.* New York: Springer Verlag, 2001.

Shayler, David J. *Skylab: America's Space Station.* New York: Springer Verlag, 2001.

Sherman, Josepha. *Deep Space Observation Satellites.* New York: Rosen Publishing Group, 2003.

Sibley, Katherine A. S. *The Cold War*. Westport, CT: Greenwood Press, 1998.

Slayton, Donald K., with Michael Cassutt. *Deke! An Autobiography*. New York: St. Martin's Press, 1995.

Sullivan, Walter. *Assault on the Unknown: The International Geophysical Year*. New York: McGraw-Hill, 1961.

Tsiolkovsky, Konstantin. *Beyond the Planet Earth*. Translated by Kenneth Syers. New York: Pergamon Press, 1960.

Voelkel, James R. *Johannes Kepler and the New Astronomy*. New York: Oxford University Press, 1999.

Walters, Helen B. *Hermann Oberth: Father of Space Travel*. Introduction by Hermann Oberth. New York: Macmillan, 1962.

Ward, Bob. *Mr. Space: The Life of Wernher von Braun*. Washington, DC: Smithsonian Institution Press, 2004.

Wills, Susan, and Steven R. Wills. *Astronomy: Looking at the Stars*. Minneapolis, MN: Oliver Press, 2001.

Winter, Frank H. *The First Golden Age of Rocketry: Congreve and Hale Rockets of the Nineteenth Century*. Washington, DC: Smithsonian Institution Press, 1990.

Wolfe, Tom. *The Right Stuff*. New York: Farrar, Straus, and Giroux, 1979.

Web Sites

"Ancient Astronomy." *Pomona College Astronomy Department*. http://www.astronomy.pomona.edu/archeo/ (accessed on September 17, 2004).

"Ancients Could Have Used Stonehenge to Predict Lunar Eclipses." *Space.com*. http://www.space.com/scienceastronomy/astronomy/stonehenge_eclipse_000119.html (accessed on September 17, 2004).

"The Apollo Program." *NASA History Office*. http://www.hq.nasa.gov/office/pao/History/apollo.html (accessed on September 17, 2004).

"The Apollo Soyuz Test Project." *NASA/Kennedy Space Center*. http://www-pao.ksc.nasa.gov/kscpao/history/astp/astp.html (accessed on September 17, 2004).

"Apollo-Soyuz Test Project." *National Aeronautics and Space Administration History Office*. http://www.hq.nasa.gov/office/pao/History/astp/index.html (accessed on September 17, 2004).

"The Apollo-Soyuz Test Project." *U.S. Centennial of Flight Commission*. http://www.centennialofflight.gov/essay/SPACEFLIGHT/ASTP/SP24.htm (accessed on September 17, 2004).

"Biographical Sketch of Dr. Wernher Von Braun." *Marshall Space Flight Center*. http://history.msfc.nasa.gov/vonbraun/index.html (accessed on September 17, 2004).

"Cassini-Huygens: Mission to Saturn and Titan." *Jet Propulsion Laboratory, California Institute of Technology.* http://saturn.jpl.nasa.gov/index.cfm (accessed on September 17, 2004).

"CGRO Science Support Center." *NASA Goddard Space Flight Center.* http://cossc.gsfc.nasa.gov/ (accessed on September 17, 2004).

"Chandra X-ray Observatory." *Harvard-Smithsonian Center for Astrophysics.* http://chandra.harvard.edu/ (accessed on September 17, 2004).

"Cold War." *CNN Interactive.* http://www.cnn.com/SPECIALS/cold.war/ (accessed on September 17, 2004).

The Cold War Museum. http://www.coldwar.org/index.html (accessed on September 17, 2004).

"The Copernican Model: A Sun-Centered Solar System." *Department of Physics and Astronomy, University of Tennessee.* http://csep10.phys.utk.edu/astr161/lect/retrograde/copernican.html (accessed on September 17, 2004).

"Curious About Astronomy? Ask an Astronomer." *Astronomy Department, Cornell University.* http://curious.astro.cornell.edu/index.php (accessed on September 17, 2004).

European Space Agency. http://www.esa.int/export/esaCP/index.html (accessed on September 17, 2004).

"Explorer Series of Spacecraft." *National Aeronautics and Space Administration Office of Policy and Plans.* http://www.hq.nasa.gov/office/pao/History/explorer.html (accessed on September 17, 2004).

"Galileo: Journey to Jupiter." *Jet Propulsion Laboratory, California Institute of Technology.* http://www2.jpl.nasa.gov/galileo/ (accessed on September 17, 2004).

"The Hubble Project." *NASA Goddard Space Flight Center.* http://hubble.nasa.gov/ (accessed on September 17, 2004).

HubbleSite. http://www.hubblesite.org/ (accessed on September 17, 2004).

"International Geophysical Year." *The National Academies.* http://www7.nationalacademies.org/archives/igy.html (accessed on September 17, 2004).

"International Space Station." *Boeing.* http://www.boeing.com/defensespace/space/spacestation/flash.html (accessed on September 17, 2004).

"International Space Station." *National Aeronautics and Space Administration.* http://spaceflight.nasa.gov/station/ (accessed on September 17, 2004).

"Kennedy Space Center: Apollo Program." *NASA/Kennedy Space Center.* http://www-pao.ksc.nasa.gov/kscpao/history/apollo/apollo.htm (accessed on September 17, 2004).

"Kennedy Space Center: Gemini Program." *NASA/Kennedy Space Center.* http://www-pao.ksc.nasa.gov/kscpao/history/gemini/gemini.htm (accessed on September 17, 2004).

"Kennedy Space Center: Mercury Program." *NASA/Kennedy Space Center.* http://www-pao.ksc.nasa.gov/history/mercury/mercury.htm (accessed on September 17, 2004).

"The Life of Konstantin Eduardovitch Tsiolkovsky." *Konstantin E. Tsiolkovsky State Museum of the History of Cosmonautics.* http://www.informatics.org/museum/tsiol.html (accessed on September 17, 2004).

"Living and Working in Space." *NASA Spacelink.* http://spacelink.nasa.gov/NASA.Projects/Human.Exploration.and.Development.of.Space/Living.and.Working.In.Space/.index.html (accessed on September 17, 2004).

"Mars Exploration Rover Mission." *Jet Propulsion Laboratory, California Institute of Technology.* http://marsrovers.jpl.nasa.gov/home/index.html (accessed on September 17, 2004).

Mir. http://www.russianspaceweb.com/mir.html (accessed on September 17, 2004).

Mount Wilson Observatory. http://www.mtwilson.edu/ (accessed on September 17, 2004).

"NASA: Robotic Explorers." *National Aeronautics and Space Administration.* http://www.nasa.gov/vision/universe/roboticexplorers/index.html (accessed on September 17, 2004).

NASA's History Office. http://www.hq.nasa.gov/office/pao/History/index.html (accessed on September 17, 2004).

National Aeronautics and Space Administration. http://www.nasa.gov/home/index.html (accessed on September 17, 2004).

National Radio Astronomy Observatory. http://www.nrao.edu/ (accessed on September 17, 2004).

"Newton's Laws of Motion." *NASA Glenn Learning Technologies Project.* http://www.grc.nasa.gov/WWW/K-12/airplane/newton.html (accessed on September 17, 2004).

"Newton's Third Law of Motion." *Physics Classroom Tutorial, Glenbrook South High School.* http://www.glenbrook.k12.il.us/gbssci/phys/Class/newtlaws/u2l4a.html (accessed on September 17, 2004).

"One Giant Leap." *CNN Interactive.* http://www.cnn.com/TECH/specials/apollo/ (accessed on September 17, 2004).

"Paranal Observatory." *European Southern Observatory.* http://www.eso.org/paranal/ (accessed on September 17, 2004).

"Project Apollo-Soyuz Drawings and Technical Diagrams." *National Aeronautics and Space Administration History Office.* http://www.hq.nasa.gov/office/pao/History/diagrams/astp/apol_soyuz.htm (accessed on September 17, 2004).

"The Race for Space: The Soviet Space Program." *University of Minnesota.* http://www1.umn.edu/scitech/assign/space/vostok_intro1.html (accessed on September 17, 2004).

"Remembering *Columbia STS-107.*" *National Aeronautics and Space Administration.* http://history.nasa.gov/columbia/index.html (accessed on September 17, 2004).

"Rocketry Through the Ages: A Timeline of Rocket History." *Marshall Space Flight Center.* http://history.msfc.nasa.gov/rocketry/index.html (accessed on September 17, 2004).

"Rockets: History and Theory." *White Sands Missile Range.* http://www.wsmr.army.mil/pao/FactSheets/rkhist.htm (accessed on September 17, 2004).

Russian Aviation and Space Agency. http://www.rosaviakosmos.ru/english/eindex.htm (accessed on September 17, 2004).

Russian/USSR spacecrafts. http://space.kursknet.ru/cosmos/english/machines/m_rus.sht (accessed on September 17, 2004).

"Skylab." *NASA/Kennedy Space Center.* http://www-pao.ksc.nasa.gov/kscpao/history/skylab/skylab.htm (accessed on September 17, 2004).

Soyuz Spacecraft. http://www.russianspaceweb.com/soyuz.html (accessed on September 17, 2004).

"Space Race." *Smithsonian National Air and Space Museum.* http://www.nasm.si.edu/exhibitions/gal114/gal114.htm (accessed on September 17, 2004).

"Space Shuttle." *NASA/Kennedy Space Center.* http://www.ksc.nasa.gov/shuttle/ (accessed on September 17, 2004).

"Space Shuttle Mission Chronology." *NASA/Kennedy Space Center.* http://www-pao.ksc.nasa.gov/kscpao/chron/chrontoc.htm (accessed on September 17, 2004).

"Spitzer Space Telescope." *California Institute of Technology.* http://www.spitzer.caltech.edu/ (accessed on September 17, 2004).

"Sputnik: The Fortieth Anniversary." *National Aeronautics and Space Administration Office of Policy and Plans.* http://www.hq.nasa.gov/office/pao/History/sputnik/ (accessed on September 17, 2004).

"Tsiolkovsky." *Russian Space Web.* http://www.russianspaceweb.com/tsiolkovsky.html (accessed on September 17, 2004).

United Nations Office for Outer Space Affairs. http://www.oosa.unvienna.org/index.html (accessed on September 17, 2004).

"Vanguard." *Naval Center for Space Technology and U.S. Naval Research Laboratory.* http://ncst-www.nrl.navy.mil/NCSTOrigin/Vanguard.html (accessed on September 17, 2004).

"Voyager: The Interstellar Mission." *Jet Propulsion Laboratory, California Institute of Technology.* http://voyager.jpl.nasa.gov/ (accessed on September 17, 2004).

"Windows to the Universe." *University Corporation for Atmospheric Research.* http://www.windows.ucar.edu/ (accessed on September 17, 2004).

W. M. Keck Observatory. http://www2.keck.hawaii.edu/ (accessed on September 17, 2004).

Index

Astrology, defined, *1:* 24
Astronaut Maneuvering Unit
(AMU), *2:* 221
Astronautics, *1:* 64, 128. *See also*
Space travel
defined, *1:* 130
Astronauts. *See also* American
astronauts
effects of acceleration on, *1:* 69
hero status of, *1:* 133
on space shuttle, *2:* 238
Astronomer(s), *1:* 1–20
Aristarchus of Samos, *1:* 26–27
Brahe, Tycho, *1:* 35–37
Copernicus, Nicolaus,
1: 33–35, 34 (ill.)
Eratosthenes, *1:* 27
Eudoxus of Cnidus, *1:* 25
Galileo (Galileo Galilei),
1: 38–41, 40 (ill.)
Hipparchus, *1:* 27–30
Kepler, Johannes, *1:* 37 (ill.),
37–38
Astronomy, *1:* 1–19, 21–43;
2: 301. *See also*
Astronomer(s); Ground-
based observatory(ies); Hub-
ble Space Telescope (HST);
Space-based observatory(ies);
Telescope(s)
ancient Greeks and, *1:* 8–9,
25–31
Arabs and, *1:* 8–9
defined, *2:* 274, 304
infrared, *2:* 292–93
as most ancient science, *1:* 8–9
radio, *2:* 287–92, 291 (ill.)
Sumerians and, *1:* 8
Astrophysicist, Spitzer, Lyman,
Jr., *2:* 308–9, 309 (ill.)
Asuka satellite observatory,
2: 328
Atacama Desert, *2:* 299
Atacama Large Millimeter Array
(ALMA), *2:* 291–92
ATDA. *See* Augmented Target
Docking Adapter (ATDA)
Atkov, Oleg, *2:* 217
Atlantis missions, 2: 252, 264,
354
Atlas-D launch vehicles, *1:* 132
ATM. *See* Apollo Telescope Mount
(ATM)

Atmosphere. *See* Earth's atmos-
phere
Atmospheric drag, *2:* 247
Atomic bomb, *1:* 96 (ill.)
defined, *1:* 88
first test, *1:* 96–97
Manhattan Project, *1:* 97–98
Auburn, Massachusetts, *1:* 72
Augmented Target Docking
Adapter (ATDA), *1:* 154–55
Aurora, *1:* 113, 114
defined, *1:* 108
Aurora 7, 1: 144–46

B

Baade, Walter, *2:* 297
Babylonians (ancient), astron-
omy of, *1:* 28
Backsight, *2:* 277
Bacon, Roger, and gunpowder
recipe, *1:* 50
Baikonur Cosmodrome, *1:* 114–15
Baikonur Space Center, *2:* 200
Ballistic missile, *1:* 110. *See also*
V-2 rockets
defined, *1:* 108, 130
Jupiter, *1:* 81
long-range, *1:* 130–31
Bamboo tubes with gunpowder,
as weapons, *1:* 48
Basic rocket equation, *1:* 66
Bassett, Charles A., II, *1:* 154
Battle of Bladensburg, *1:* 55–56
Battle of Fort McHenry (Balti-
more, Maryland), *1:* 45–47,
46 (ill.), 56
Battle of Leipzig (Battle of the
Nations), *1:* 55
Bazooka, *1:* 71
Beagle 2 Mars lander, *2:* 351
Bean, Alan L., *1:* 179; *2:* 220
Bellifortis (War Fortifications) (von
Eichstadt), *1:* 51
Belyayev, Pavel, *1:* 149
Bends, *2:* 200
defined, *2:* 190
Beregovoi, Georgi T., *1:* 175
Berkner, Lloyd, *1:* 113
Berlin Airlift, *1:* 101
Beyond the Planet Earth (Tsi-
olkovsky), *1:* 69; *2:* 209

Cultures, and names of constellations, *1:* 14–16
Cunningham, R. Walter, *1:* 174
CXO. *See* Compton Gamma Ray Observatory (CXO)
Cygnus (constellation), *1:* 9, 11

D

Dark matter, defined, *2:* 274, 304
Dark skies, for ground-based observatory, *2:* 287
Das Problem der Befahrung des Weltraums: Der Raketen-motor (The Problem of Space Travel: The Rocket Motor) (Potocnik), *2:* 211
Day, *1:* 10
  sidereal, *1:* 4, 10
  solar, *1:* 4, 10, 11
The Day the Earth Stood Still (film), *1:* 108
De caelo (On the Heavens) (Aristotle), *1:* 25
de Gaulle, Charles, *2:* 196
De la pirotechnia (On Working with Fire) (Biringuccio), *1:* 52–53
De mirabilibus mundi (On the Wonders of the World) (Albertus Magnus), *1:* 50
De nova stella (Concerning a New Star) (Brahe), *1:* 36
De Revolutionibus Orbium Coelestium (Revolution of the Heavenly Spheres) (Copernicus), *1:* 34
Decent module (Apollo-Soyuz), *2:* 188 (ill.)
Declaration on Liberated Europe, *1:* 95
Deep Space 1 space probe, *2:* 361
Delta-winged orbiter, of space shuttle, *2:* 239–41, 240 (ill.), 243 (ill.), 244–48
Democracy, *1:* 90
  defined, *1:* 88–89
Deneb (star), *1:* 9, 11
Descartes Highlands, *1:* 182
Destiny control module, of International Space Station, *2:* 233

Détente, *2:* 190, 194–96
Dialogo Galilei linceo . . . sopra i due massimi sistemi del mondo (Dialogue on the Two Chief Systems of the World) (Galileo), *1:* 40
Dialogue on the Two Chief Systems of the World (Dialogo Galilei linceo . . . sopra i due massimi sistemi del mondo) (Galileo), *1:* 40
Die Rakete zu den Planeträumen (The Rocket into Planetary Space) (Oberth), *1:* 75; *2:* 210
Discoverer 14, *1:* 125
Discovery missions, *2:* 252, 262, 264, 317
Disney, Walt, *1:* 110; *2:* 211
DM. *See* Docking Module (DM), of Apollo-Soyuz test project
Dobrovolsky, Georgy, *1:* 175; *2:* 192, 213
Docking module (Apollo-Soyuz), *2:* 188 (ill.), 198–200
Docking system, defined, *2:* 190
"Dog days of summer," *1:* 11–13
Dog Star (Sirius), *1:* 6, 11–13
Dollard, John, *2:* 283
Draco (constellation), *1:* 10, 14
"Dreams of the Earth and Sky and the Effects of Universal Gravitation" (Tsiolkovsky), *1:* 65–66
Dryden, Hugh L., *1:* 126 (ill.)
Duke, Charles M., Jr., *1:* 182
Dwarf galaxies, *2:* 293

E

EADS Phoenix space shuttle, *2:* 267
Eagle lunar landing module, *1:* 178
Earth
  ancient Greeks and, *1:* 25–31
  axis of, *1:* 9–10, 28. *See also Precession*
  circumference of, *1:* 27
  distance from Moon, *1:* 29
  distance from Sun, *1:* 5–6
  magnetic field, *1:* 120
  movement through space, *1:* 9–14

K

Kaluga, Russia, *1:* 63
"Kaputnik," *1:* 120
Kazakhstan, *1:* 99, 114
Keck Telescopes, *2:* 297–98, 298 (ill.)
Kennedy, John F., *1:* 137, 145 (ill.), 164 (ill.); *2:* 195
  vows to put man on the moon, *1:* 160, 162; *2:* 211
Kennedy Space Center, *1:* 167; *2:* 201, 244, 247
Kepler, Johannes, *1:* 37 (ill.), 37–38, 43; *2:* 282
Kerwin, Joseph P., *2:* 220
Key, Francis Scott, *1:* 45–47, 46 (ill.), 56
Khrunov, Yevgeny, *1:* 177
Khrushchev, Nikita, *1:* 119, 119 (ill.), 120, 135, 147
Kissinger, Henry A., *2:* 196
Kizim, Leonid, *2:* 217
Klimuk, Pyotr, *2:* 215
Komarov, Vladimir, *1:* 149, 174, 175
Kopernik, Nicolause. *See* Copernicus, Nicolaus
Korolev, Sergei, *1:* 116, 119, 165
Kosygin, Alexei, *2:* 196–97
Krikalev, Sergei, *2:* 234
Kristall module, *2:* 224–25
Kubasov, Valeri, *1:* 179; *2:* 198, 199 (ill.)
Kummersdorf, Germany, *1:* 79
Kvant 1 module, of *Mir, 2:* 224
Kvant 2 module, of *Mir, 2:* 224

L

L-1 spacecraft, *1:* 165–67; *2:* 192
L-3 program, *1:* 165–67; *2:* 192
Lacaille, Nicolas Louis de, *1:* 8
Laika (first animal in space), *1:* 117
Landing, hard, defined, *2:* 335
Landing, soft, defined, *2:* 335
Lang, Fritz, *1:* 76
Langley Aeronautical Laboratory, *1:* 125
Languages, in Apollo-Soyuz test project, *2:* 198
Large Space Telescope (LST) project, *2:* 315
Law(s) of motion
  Newton, Isaac, *1:* 42–43, 53
  third, *1:* 66–67
Laws of planetary motion, Kepler, Johannes, *1:* 37–38
Lead-gold alloy, for Apollo-Soyuz test project, *2:* 203
Lenin, Vladimir I., *1:* 89–90, 90 (ill.), 130
Lens(es), *2:* 278–79
  achromatic, *2:* 283
  concave, *2:* 274
  convex, *2:* 274
Leo (constellation), *1:* 11
Leonov, Aleksei, *1:* 148, 148 (ill.), 149, 163; *2:* 198–205, 199 (ill.)
Leviathan of Parsonstown, *2:* 284
Lewis Flight Propulsion Laboratory, *1:* 125
Liberty Bell 7 (Mercury spacecraft), *1:* 143
Library of Alexandria, Egypt, *1:* 23–24
  Eratosthenes, *1:* 27
Light, *2:* 301
  extreme ultraviolet, *2:* 312
  speed of, *1:* 4–5; *2:* 299
  and time, *1:* 5–6
  visible, *2:* 273, 278, 301
Light rays, distortion of, *2:* 287
Light-year, defined, *1:* 4; *2:* 274–75, 304
Lindbergh, Charles, *1:* 72
Linenger, Jerry, *2:* 225
Lipmann, Walter, *1:* 87
Lippershey, Hans, *1:* 39; *2:* 279 (ill.), 279–80
Liquid hydrocarbon, *1:* 67
Liquid hydrogen, *1:* 67; *2:* 242
Liquid oxygen, *1:* 67; *2:* 242
Liquid-fuel rocket, defined, *1:* 48
"Liquid-propellant Rocket Development" (Goddard), *1:* 73
Liquid-propellant rockets, *1:* 67–69, 68 (ill.)
  in engines of space shuttle, *2:* 240 (ill.), 242, 243 (ill.)
  first launch of, *1:* 72
  Oberth's work on, *1:* 74, 76, 77

vs. solid-propellant rockets, *1:* 68 (ill.)

Llactapata, *2:* 277

LM. *See* Lunar module *Eagle*

Long-range ballistic missiles, *1:* 130

Lousma, Jack R., *2:* 220

Lovelace, W. Randolph, II, *1:* 140

Lovell, James Jr., *1:* 153, 157, 176 (ill.), 176–77, 180–81

Low, George, *2:* 196

LRV. *See* Lunar Roving Vehicle (LRV)

Lubbock, Sir John William, *1:* 111

Lucid, Shannon, *2:* 225, 226, 227 (ill.)
   spends six months in space, *2:* 225

Luna space probe series (Soviet), *2:* 336–38
   Luna 1, 1: 162; *2:* 189, 333, 336
   Luna 2, 2: 189, 336–38
   Luna 3, 1: 69; *2:* 189, 337
   Luna 9, 2: 337
   Luna 13, 2: 337
   Luna 15, 2: 337
   Luna 16, 2: 337
   Luna 17, 2: 337
   Luna 21, 2: 337
   Luna 24, 2: 337

Lunakhod Moon rovers, *2:* 337

Lunar eclipses, *1:* 19, 25, 26

Lunar landings, *1:* 178–84; *2:* 189. *See also* Project Apollo

Lunar module *Eagle, 1:* 167–68, 170–71

Lunar Orbiter space probe series (U.S.), *2:* 339–40
   Lunar Orbiter 3 space probe, *2:* 339

Lunar Prospector, 2: 340–41

Lunar Roving Vehicle (LRV), *1:* 181

Luther, Martin, *1:* 34–35

Lyra (constellation), *1:* 11

M

Macedon, *1:* 21–23

Machu Picchu, *2:* 277

Magellan spacecraft, *2:* 264, 346

Magnetic field, *1:* 120
   defined, *2:* 335

Magnetism, defined, *1:* 108

Magnetosphere, defined, *2:* 335

Magnifying glasses, *2:* 278

al-Majisti (Almagest; The Greatest; The Mathematical Compilation; He mathematike syntaxis) (Ptolemy), *1:* 30

Malyshev, Yuri, *2:* 215 (ill.)

"Man and the Moon" (TV show), *1:* 110

"Man in Space" (TV show), *1:* 110

Manhattan Project, *1:* 97–98

Manned Maneuvering Unit (MMU), *2:* 254

Manned spaceflight, *1:* 128–59; *2:* 189
   Soviet program for, *1:* 117, 133–37
   U.S. program for, *1:* 127, 139–46; *2:* 189, 248–49, 318–19. *See also* Project Apollo; Project Gemini; Project Mercury; Space shuttle(s)

Mao Zedong, *1:* 101

Mariner space probes, *2:* 342–43, 343 (ill.)
   Mariner 2, 2: 344
   Mariner 3, 2: 347
   Mariner 4, 2: 347–48
   Mariner 6 and 7, 2: 348
   Mariner 9, 2: 348–49
   Mariner 10, 2: 342, 343 (ill.)

Mars, *2:* 347–53
   Kepler, Johannes, and, *1:* 38

Mars 1–5 space probes, *2:* 347

"Mars and Beyond" (TV show), *1:* 110

Mars Climate Orbiter space probe, *2:* 350–51

Mars exploration
   Mars Exploration Program, *2:* 349, 351–53
   Mars Global Surveyor, *2:* 349, 352 (ill.)
   Rover Mission, *2:* 351–53

Mars Express space probe, *2:* 351

Mars Global Surveyor space probe, *2:* 349, 352 (ill.)

as space junk, *2:* 360
Velas (military), *2:* 319
Saturn, *2:* 356–58
Saturn 5 rocket, *2:* 189, 217
    for *Apollo 11, 1:* 168 (ill.),
        168–70, 169 (ill.)
    von Braun, Wernher, and, *1:* 83
Saturn IB rocket, *1:* 169, 174
Saucepan (constellation), *1:* 15
Savery, Thomas, *1:* 47
Savitskaya, Svetlana, *1:* 139
Schirra, Walter M., Jr., *1:* 141,
        146, 153, 174; *2:* 200
Schmitt, Harrison H., *1:* 182
Schweickart, Russell L., *1:* 177
Science and Engineering Re-
        search Council (UK), *2:* 310
Science fiction, *1:* 106–9
    Cyrano de Bergerac, Savinien
        de, *1:* 61
    enjoyed by Hermann Oberth,
        1: 74
    inspires Konstantin Tsiolkovsky,
        1: 69
    inspires Robert Goddard, *1:* 70
    inspires space stations, *2:* 209
    Verne, Jules, *1:* 61–62, 64, 70
    War of the Worlds broadcast,
        1: 100
Science fiction writers
    Hale, Edward Everett, *2:* 209
    Tsiolkovsky, Konstantin, *1:* 69;
        2: 209
    Wells, H. G., *1:* 61–62, 64, 70
Science, Hellenistic, *1:* 24–25
Science, natural, defined,
        1: 108–9
Science, physical, defined,
        1: 108–9
Science Satellite 1 (Sci-Sat),
        2: 327–28
Science Satellite 1 (SciSat-1),
        2: 327–28
Scientific experiments. *See* Exper-
        iments
Scientific observation
    by ancient Greeks, *1:* 26
    by space probes, *2:* 333
Scientist-astronauts, on *Skylab*
        space station, *2:* 218
Scintillation, stellar, *1:* 3–4
    defined, *1:* 4
SciSat-1 (Science Satellite 1),
        2: 327–28

Scobee, Francis (Dick), *1:* 175;
        2: 255, 256 (ill.)
Scooter (Neptune), *2:* 360
Scott, David R., *1:* 154, 177, 181,
        182
Sea of Serenity, *1:* 182; *2:* 336
Sea of Tranquility, *1:* 178
Seasons, *1:* 10
    and position of stars, *1:* 11–14,
        16
Sedna planetoid, *2:* 294
See, Elliot M., *1:* 154
Sensors, on space probes, *2:* 333
September 11, 2001, *2:* 265–56
Service module (Apollo space-
        craft), *1:* 167
Service module (Apollo-Soyuz),
        2: 188 (ill.)
Service Propulsion System (SPS),
        1: 175
Sevastyanov, Vitali, *2:* 215
Shatalov, Vladimir A., *1:* 177, 179
Shenzhou 5 spacecraft, *1:* 156
Shepard, Alan, *1:* 141, 143,
        145 (ill.), 162, 181; *2:* 189
Shepherd, William M. "Bill,"
        2: 234
Shoemaker, Eugene M., *2:* 341,
        362
Shoemaker-Levy 9 comet, *2:* 317,
        341
Shonin, Georgi S., *1:* 179
Shuttle-*Mir* missions, *2:* 226
Shuttleworth, Mark, *2:* 234
Sidereal day, *1:* 10
    defined, *1:* 4
Sidereus Nuncius (Starry Messenger)
        (Galileo), *1:* 40; *2:* 280–81
Siemienowicz, Kazimierz, rock-
        etry theory, *1:* 53–54
Sight, in stargazing, *2:* 277
Sight lines, in ancient observato-
        ries, *2:* 277
Sigma 7, 1: 146
Sirius (Dog Star), *1:* 6, 11–13
SIRTF. *See* Space Infrared Tele-
        scope Facility (SIRTF)
67/PChurymov-Gerasimenko
        comet, *2:* 362
Sky
    best, for ground-based observa-
        tory, *2:* 287
    study of, *2:* 271–76. *See also*
        Astronomy

SS6 rocket, *1:* 119

Stafford, Thomas P., *1:* 153, 154; *2:* 198–205, 199 (ill.)

Staging, in rocketry, *1:* 68–69, 98

Stalin, Joseph, *1:* 86 (ill.), 89, 92–97, 94 (ill.), 116

Standing stones (Stonehenge), *1:* 16 (ill.), 17

Star City, *2:* 234

Star Trek (TV show), *2:* 251

"Star war events," *1:* 19

Stardust space probe, *2:* 361

Stargazers (ancient), name constellations, *1:* 14–16

Starry Messenger (Sidereus Nunciusr) (Galileo), *1:* 40; *2:* 280–81

Stars, *1:* 1–20. *See also* Constellations; specific stars
    binary, defined, *2:* 304
    composition of, *1:* 1
    defined, *1:* 1, 4
    exploding, *2:* 311–12
    formation of, *2:* 293
    generate light, *1:* 3–4
    groups of, *1:* 7
    Hipparchus's catalog of, *1:* 28
    infrared observation of, *2:* 326
    neutron, defined, *2:* 275, 305
    scale of magnitude (Hipparchus), *1:* 28
    twinkling of, *1:* 3–4

"The Star-Spangled Banner," *1:* 45–47, 46 (ill.), 56

Steam engine, *1:* 47

Stellar nurseries, *2:* 293

Stellar scintillation, *1:* 3–4
    defined, *1:* 4

Stellar wind, defined, *2:* 305

Stickless rockets, *1:* 56

Stjerneborg Observatory *(Castle of the Stars)*, *1:* 36

Stock Market Crash (1929), *1:* 79

Stonehenge, *1:* 16 (ill.), 16–17

Storytelling, and naming constellations, *1:* 14–15

Strategic Arms Limitation Treaty (SALT), *2:* 196

STS. *See* Space Transportation System (STS)

STS-1 shuttle flight, *2:* 252

STS-5 shuttle flight, *2:* 252–53

STS-26 shuttle flight, *2:* 262

STS-30 shuttle flight, *2:* 264

STS-31 shuttle flight, *2:* 264

STS-34 shuttle flight, *2:* 264

STS-37 shuttle flight, *2:* 264

STS-41 shuttle flight, *2:* 264

STS-50 shuttle flight, *2:* 263–64

STS-51L. *See Challenger*

STS-57 shuttle flight, *2:* 265

STS-93 shuttle flight, *2:* 264

STS-107 shuttle flight, *2:* 266. *See also Columbia*

Stuart, J.E.B., *1:* 57

S-turn maneuvers, of shuttle orbiter, *2:* 247

Subaru Telescope, *2:* 297

Suborbital flights, Mercury project, *1:* 142–43

Sumerians, and constellations, *1:* 8

Summer solstice, *1:* 15; *2:* 277
    Eratosthenes and, *1:* 27
    Stonehenge and, *1:* 16 (ill.), 17

Summer triangle, *1:* 11

Sun, *2:* 363–64
    ancient Greeks and, *1:* 25–31
    Aristarchus of Samos, *1:* 26–27
    El Caracol and, *1:* 18 (ill.), 19
    distance from Earth, *1:* 5–6
    Earth's orbit around, *1:* 11
    infrared observation of, *2:* 326
    in Mayan life, *1:* 19
    patterns observed by ancient cultures, *1:* 15–16
    star nearest to, *1:* 6
    Stonehenge and, *1:* 16 (ill.), 17
    and ultraviolet radiation, *2:* 306
    Ulysses probe to, *2:* 264, 363–64

Sun-centered model of planetary motion. *See* Heliocentric (Sun-centered) model of planetary motion

Sunspot, *1:* 113
    defined, *1:* 108–9; *2:* 210, 275, 305

Supernova
    Brahe, Tycho, *1:* 36
    Chandrasekhar limit, *2:* 322
    defined, *1:* 24; *2:* 275, 305
    first identification of, *2:* 311–12